BEEKEEPING

VOLUME 1

RON BROWN OBE B.Sc

from the archives of the Beekeepers Quarterly for a new generation of beekeepers

EDITED BY JOHN PHIPPS 2014

PHOTO
RON BROWN (Chosen by Rosemary Brown)

First published by Northern Bee Books, 2014.
First Edition.

Northern Bee Books,
Scout Bottom Farm,
Mytholmroyd,
Hebden Bridge,
West Yorkshire HX7 5JS.

http://www.northernbeebooks.co.uk

ISBN 978-1-908904-64-5

Printed & Bound by
Lightning Source, an INGRAM Content Company.
http://www.lightningsource.com

CONTENTS

Foreword By John Phipps 9

1. LOOK, LISTEN AND LEARN 11

   ## BKQ 36 Winter 1993/4
   Finding out What's Going on in Your Hive Without Taking off the Roof 11
   Pollen Loads Going In 11
   Fanning Bees at Entrance 13
   Fanning With Tails Up 14
   Fanning in Very hot Weather, Tails Down 14
   Tiny wax Platelets Seen on the Alighting Board 14
   Fine Dust of Wax Seen on Alighting Board During Winter ... 15
   Larger Crumbs of Wax 15
   Larger Curved Pieces of Wax 15
   Hard Grey Pollen Pellets 17
   Chalk Brood Mummies 17

   ## BKQ 37 Spring 1994
   Dead Bees on the Ground 18
   Brown Smears on Hive Front 19
   Bumblebees 20
   Acarine Testing 20
   Drones Flying Freely 21
   Honey Flow 21
   Swarming Bees 22
   Swarm-watching 22
   White Grubs Thrown Out 24

Sounds 25
Propolis Collecting 26

BKQ 38 Summer 1994

Wasps Trying to Enter 27
Wasps Entering Freely 27
Ants 28
Drones Being Ejected by the Bees 29
Very Small Drones 29
Robbing Bees 30
Death of Flying Bees (Summer) 31
Cluster of Bees on Hive Front 31
Gaps Between Supers 32
Hefting Hives 32

BKQ 39 Autumn 1994

Smells at Feeder Hole (Winter) 34
Apparently Lifeless Bees on the Ground 35
Slugs 36
Woodpeckers 36
Temperature and Flight 36
Instinct or Intelligence? 37

2. **ECONOMICAL BEEKEEPING** 40

BKQ 41 Spring 1995

Equipment:
Queen Excluders 40
Honey Liquefying Cabinet 42
Double Nucleus From Old WBC Brood Box 44
Solar Wax Extractor 44
Wax From Old Black Combs 47

BKQ 42 Summer 1995

Warming Honey Supers 47
Economical Use of Swarms 51
Reinforcing 52

CONTENTS

    Induced Supersedure 53
    Wax Saver 54

BKQ 43 Autumn 1995
    Maintenance of Equipment 56
    Hive Products 57
    Harvesting Pollen 58
    Pollen 59
    Homogenising Pollen/Honey 59
    Harvesting Propolis 60
    Mead from Cappings 61
    Package Bees 63

3. A TOP-BAR HIVE 64

BKQ 44 Winter 1996/7
    Natural Material 64
    Comb Inspection 67
    Honey Harvest 67
    Feeding 68
    Roof 68

4. SWARMS ARE PRECIOUS: WHY NOT TRY A BAIT HIVE AND SEE WHAT HAPPENS? 69

BKQ 57 Spring 1999
    Why not Prepare a Bait Hive and see What Happens? 69
    Using a Swarm to Plump up a Colony 71

5. VARROA: A REVIEW 73

BKQ 30 Summer 1992
    Arrival in Europe 73
    Cross-country Spread 74
    Experience Elsewhere 75
    Four Years to Kill 76
    Control by Management 78

Chemical Residues 79
Long-term Research 79
The Way Ahead 81
The Next Cloud on the Horizon? 82

BKQ 31 Autumn 1992
VARROA: Your Questions Answered 83

VARROA: Varroa Mite Knock-down Rates 87

BKQ 32 Winter 1992/3
Deductions from Data (Hive G) 89
Assumptions (Hive G) 89
Calculations (Hive G) 89

VARROA: Development of Varroa Mite Resistance to Synthetic Pyrethroids 89

BKQ 49 Spring 1997
Varroa: Thymol v Varroa 91

BKQ 60 Spring 2000
1. Thymol 91
2. What is Thymol? 92
3. Composition of Italian Sachets of Apilife Var: 92
4. Thymol Dispersion 92
5. Vaporisation of Thymol Crystals 93
6. Use in Winter Feed 93
7. Use in Summer 94
8. Some Practical Details 94
9. My Current Practice 95

6. **BROTHER ADAM, AN APPRECIATION** 97
BKQ Autumn 1996

CONTENTS

7. THE KILLER BEES OF BRAZIL 99

The Beekeepers Annual 1983
- Background 99
- Accidental Queen Release 99
- The Florianopolis Conference 100
- The Tour 101
- Why so Aggressive? 102
- What of the Future? 103

8. RON BROWN OBE BSC 105
BKQ 104 Summer 2011

BKQ 86 November 2006
- The man and his Books an Appreciation by Val Phipps 105
- Key Dates 107
- April 1953 107
- May 1965 107
- 1975 -1982 108
- 1981 108

THE BOOKS: Autobiographical Books: 108
- All round the compass with RAF Coastal Command 108
- Ex Africa 108

BEEKEEPING BOOKS 109

FOREWORD
By John Phipps, Editor, June 2014

Since its first publication in 1984, The Beekeepers Quarterly has attracted some of the best writers from all parts of the world. Many of the contributors have had a specific interest in a particular sphere of beekeeping and have brought expert knowledge to our pages, thus keeping our readers up-to-date with the latest advances in beekeeping science, technology and practice.

Most of our writers have had many, many years of beekeeping experience behind them which has always allowed them to see emerging problems in the context of past events, giving them an overview which cannot be surpassed by recent writers.

Of importance, too, is the fact that their perspective is not only historical, but also global, for they have concerned themselves with what is happening on an international scale which has given them a decided advantage over those who have a more parochial interest in the craft.

The Beekeepers Quarterly has always been concerned to present aspects of beekeeping in the widest possible way and this has been reflected in the work of our team of writers, many of whom who have been connected with the magazine for decades.

Much of the material written over the years is still of great value and relevance to beekeepers who have recently started out in beekeeping. So, in order to make this material available, we are publishing volumes of past articles written by our major contributors. Inevitably, there are areas where a very minor amount of the information has changed - this is particularly so regarding beekeeping pests and their treatments and the reader is therefore encouraged to seek the latest advice on recommended treatments.

This first volume is devoted to pieces written for us by Ron Brown, who had extensive knowledge of beekeeping abroad as well as in the UK, and who had published some of the best books on practical beekeeping for beekeepers of all abilities. However, some of the material within this volume is not contained in his books, so we are pleased to be able to make it available for those who were not subscribers to earlier issues of the BKQ.

SECTION 1.
LOOK, LISTEN & LEARN

BKQ 36
Winter 1993/4

Finding out What's Going on in Your Hive Without Taking off the Roof

In beekeeping, as in every other human activity, there is a factor of experience which can sometimes enable a keen observer to tell at a glance what is happening inside a hive, without the hassle of smoking, opening up and pulling out frames of bees. There is no complete substitute for forty years of 'hands-on' experience, but the following notes may help the reader to interpret what he sees and to become 'experienced' more quickly. Your bees (and your close neighbours) may also gain. Observation is easier if a hive has an alighting board, and the diagram illustrates a simple pattern, easily made and tacked on just below the entrance and removed by simply pulling it off if a hive has to be moved. The only other apparatus needed is a stool - I use a strong plastic beer case with a carpet square (18" x 18") top, plus a hand-held simple magnifying glass.

Pollen Loads Going In

This is a good sign, and abundant pollen loads indicate a laying queen with plenty of larvae being fed, to build up a strong force of worker bees. For hundreds of years beekeepers have written about this and it is still true today. Usually

PLAN

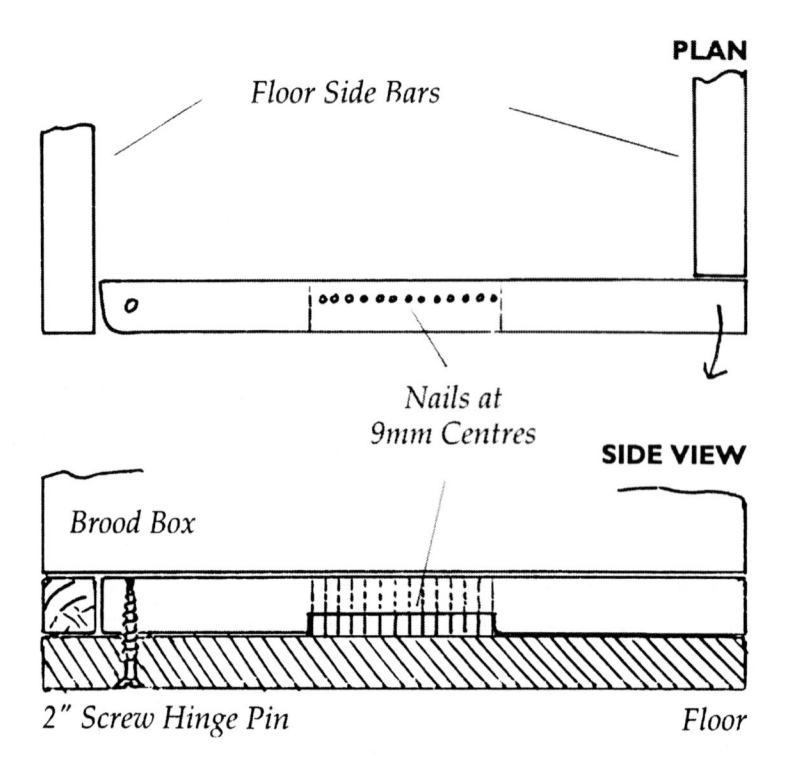

Floor Side Bars

Nails at 9mm Centres

SIDE VIEW

Brood Box

2" Screw Hinge Pin   Floor

Image: Before fitting the new entrance block a 7/8" piece of the right hand batten of the floorboard is cut away. A 7/8" block of wood is then cut to the right length for the type of hive used and in the centre a strip 4" x 1/4" is then cut away from the underside. Thin 7/8" nails are then inserted at 9mm intervals through the top of the entrance block along the part through which the entrance has been cut. The left hand edge of the block is shaved away slightly to form a curve to allow the entrance to be hinged open, when fastened with a 2" screw through the floorboard and into the block. Compared with other types of entrance/mouse guard this can be used throughout the year and swivelled open when the colony becomes more populous or if varroa sheets need to be inserted or removed. Pollen is not likely to be scraped off the legs of bees if this type of mouse guard is used.

one will see this from February onwards, often when it is too cold to open up anyway. The size of pollen pellets on bee hind legs can be an indication of the amount of forage in the area. Large pellets averaging 20-25 mg on each corbicula are good: small pellets may just be young bees learning their trade, or pollen collected by accident. To be really up-to-date, you need a copy of Dorothy Hodges' "Pollen Loads of the Honeybee", which has six pages of coloured illustrations of 120 different pollens, from Laurestinus *(Viburnum tinus)* in January to Ivy *(Hedera helix)* in October and early November. Dates of usual appearance will vary from year to year and from one part of Britain to another, but the order in which they appear will not vary much. Laurestinus pollen is white; dandelion, a golden yellow; bluebell a slate-blue; horse chestnut, brick red; clover putty-grey; meadowsweet, pale green; ivy does vary sometimes, from a pale buff yellow to a brighter yellow. I have spent many happy hours sitting by a hive with "Pollen Loads" on my knees and bees alighting on my arm to rest for a while under close inspection; I have even had a bee carrying horse chestnut pollen land on the open page and walk over the very illustration, and been genuinely unaware until that moment that a chestnut tree was already in blossom.

Fanning Bees at Entrance

- with tails down, late at night and early in the morning, as well as during the day, is a very good sign. Accompanied by the loud hum of many wings at work inside the hive and also a strong smell of nectar, this is the evidence of a good honey flow by day and thin nectar being evaporated down into good, ripe honey 24 hours a day. There may well be a wet patch of condensed water in the cool, early morning on the alighting board opposite the entrance. The smell of the nectar will indicate what the bees are working, like sycamore in May, or ivy in October. In the early morning, with no other sounds, one can hear the roar of fanning bees a dozen yards away.

Fanning With Tails Up

- and scent glands exposed conveys a different meaning altogether. This means that the bees are blowing out a scent to act as a beacon, with the message, "Come on in, this is your hive". Close inspection will show a moist ellipse of bare tissue between segments on the backs of bees, near tails, and one can identify the Nasonov pheromone, with its smell of lemon balm. This navigational aid will be noticed if a hive has been moved and the bees need to draw others into a new site, but it is most clearly seen when hiving a swarm, when perhaps hundreds of bees will be fanning.

I have also noticed it when there is a nursery flight of young bees learning to find their way home, or when a nucleus has a queen out on a mating flight. The scent is not specific to any particular colony. The interpretation depends very much on the situation, but always indicates that flying bees are picking up the chemical message on their antennae from several feet away, and homing in on it.

Fanning in Very hot Weather, Tails Down

- With no honey Flow, will indicate that the bees are collecting and evaporating water just to keep the hive cool. At a time like this they will be seen collecting water from a local tap, fishpond or swimming bath.

Tiny wax Platelets Seen on the Alighting Board

- Or being blown out of the hive entrance, usually in early summer, indicate a surplus of young wax-secreting bees with no work to do, or if from a recently hived swarm, that wax is being produced on a large scale and some dropped. In early summer, some wax-building work should always be given, such as one or two frames of foundation to be drawn out. In nature, young bees are programmed to produce wax to build new combs and if beekeepers go on using the same old brood frames and only put on supers of drawn comb, the bees find no useful outlet for their new wax and either drop it on the hive floor or build comb wherever there is any room. It pays to 'go along with the bees' and let them do some comb-building, perhaps to draw out a new super from foundation, or even produce honeycombs for sale.

Fine Dust of Wax Seen on Alighting Board During Winter

- Even when bees are not flying perhaps for weeks at a time, is a good sign. It comes from normal uncapping of food cells slowly and steadily as the bees need it, and indicates that all is well inside.

Larger Crumbs of Wax

- Up to a tenth of an inch in size suggests robbing, either by bees or wasps. the robbers are in a hurry and tear down cappings and sometimes the edges of the cells. These can be seen in spring or autumn, possibly in summer, but not usually in winter. If first noticed in winter, it probably happened late in autumn.

Larger Curved Pieces of Wax

- up to half an inch, but usually about a quarter, found only in winter, are a sure sign of mice in the hive. Very often black droppings will be seen as well. Mice will enter unprotected hives in late autumn, at a time when the bees have clustered on central combs. They will eat a hole in outer store combs and bring in grass and leaves to build a nest; this will protect them from the bees, should the weather turn warmer again. What should one do, on noticing this in January, say? Orthodox wisdom says that one must not open up a brood box at this time of year, but if the bees are going to die anyway,

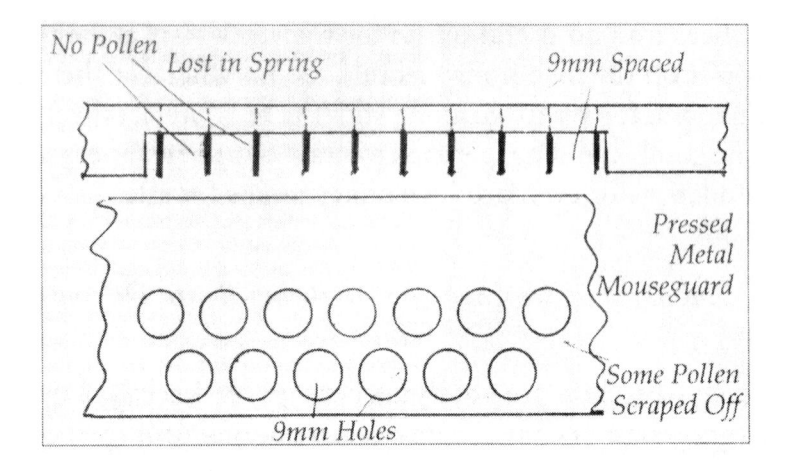

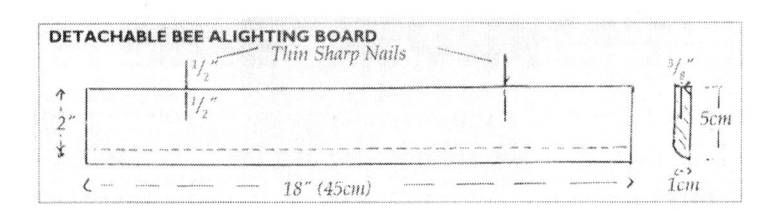

Image: Detachable Bee Alighting Board. This is made from a wooden strip measuring 18" x 2" X 3/8" (45 x 15 x 1cm). Holes are drilled in the side of the board to hold the blunt half of thin nails with heads cut off. The board is gently tapped into the hive at floor level, below the entrance.

Whilst being a great help for the observation of bees, it also helps heavily laden bees to land.

perhaps one should for this condition. I have waited for a relatively mild day and then transferred unaffected combs into a new box, three or four at a time, plus bees, leaving the other frames plus mouse nest in the old box. Usually the mice occupy three or four combs at one side only, but the dirt and smell of them are so repugnant to the bees that they lose heart and need a fresh start, with at least two combs of food to replace the mousy frames. The old box can then be taken away and the mouse nest and occupants removed. The actual comb beneath the nest will sometimes be so contaminated with mouse urine that the whole comb has to be cut out and burnt, but perhaps adjacent combs, with only one face gnawed down can be sterilised with 80% acetic acid along with others in routine treatment in April, and re-used.

Prevention is better than cure of course, and the message is that a mouse guard should have been fitted last October, unless a special entrance is in use, like that shown in the drawing, for example.

Hard Grey Pollen Pellets

- the size of a cell may sometimes be noted on the alighting board in March and April. They could be mistaken for chalk brood, but if crushed, will break up and show striations or layers, sometimes with a trace of colour left. The fact that they are being thrown out is good, because it shows that the brood nest is expanding and the queen is laying well. However, this has given a great deal of unnecessary hard work to the bees, and demonstrates that there were not enough active bees in the autumn to preserve and cap those cells of pollen to prevent mould. Perhaps an additional autumn feed of syrup could have prevented this.

Chalk Brood Mummies

These do not crumble into layers as the mouldy pollen does, but sometimes the mummified larvae do take the shape of the cell and appear similar to the stale pollen just referred to. Usually they are flatter and still recognisable as poorly developed pupae. There is no particular treatment for this condition, except to re-queen and hope that new blood will have more natural resistance. Often, traces of chalk brood can be seen at one summer inspection and can be miss-ing three weeks later. It is associated with weak colonies, perhaps suffering from some other condition; strong colonies, well-fed and on good combs, are less likely to have chalk brood. As Shakespeare said "When troubles come, they come not single spies but in battalions", and this is as true of bees as of humans. Since the arrival of varroa in S. England, there have been many reports that chalk brood is more often found in weakened colonies. In fact, if you see a lot more chalk brood than usual, you may have previously undetected varroa!

BKQ 37
Spring 1994

Dead Bees on the Ground

When your hive is in the garden and glanced at most days, you will inevitably notice some dead bees on the ground under the entrance. Every year I get phone calls from kind-hearted beekeepers who are worried by this, usually on a cold day in spring. After checking that the amount is not excessive (more than half a cupful) I usually explain that bees (like humans) are mortal. Most die 'with their boots on' far from the hive and are never noticed, but at the end of the winter many die of old age 'in bed' and have to be disposed of. If flying is possible, then undertaker bees will fly out with them and drop them 20 or 30 yards away un-noticed. Failing this their bodies are just pushed out of the hive. I usually go on to say that a typical hive population in September is about the same as that of Torquay (about 40,000 but by February is down to 10,000, so that statistically 30,000 (or more) die in roughly 150 days, i.e. about 200 a day. Humans do live longer, but bookings at the local cemetery are also rather congested in February and March. If your hive stand is on a smooth concrete or tarmac surface you are more likely to notice the natural mortality; if on a lawn you may not even be aware of it. If the quantity of dead bees is much greater than half a cupful mentioned, you may have a disease problem. The presence of K-wingers, bees with swollen abdomens or, on milder days bees walking away from the hive on the ground or crawling up grass stems, could indicate acarine. (See Hive-side testing technique).

Brown Smears on Hive Front

Oh dear! Dysentery? Not necessarily, but it will be excreta from bees which have crawled out of the hive, or flown out and landed back almost at once out of respect for your white shirt on the wash-ing line. The interpretation depends on the circumstances, which can vary, for example:

(a) On a cold day in spring, when the queen is laying well and nurse bees are eating large quantities of stored pollen in order to manufacture brood food via their hypopharyngeal glands, they rapidly fill up their recta with empty husks of pollen grains, occupying the same volume as when full. This healthy activity necessitates regular bowel movements, and on a cold day they can't fly very far, perhaps only as far as the highly polished roof of your neighbour's new car! Or sometimes on the hive front.

(b) If your hive was taken to the moor last autumn it will have heather honey stored in the brood chamber, and in colder areas of Britain this may result in brown smears. Heather honey contains almost 2% of protein in the actual honey, quite apart from that in stored pollen. This is why they build up so well in the spring, with no problems in the south-west where cleansing flights are usually possible at intervals throughout most winters. In north-east Britain, where bees are more often confined by colder weather heather honey can cause brown smears. Any protein food produces some solid residues, unlike normal honey (or sugar syrup) which is metabolised only to water vapour and carbon dioxide. So do what Brother Adam does: feed a gallon or more of syrup to every stock brought back from the moor early in September. Last in, first out; the pure carbohydrate is used up in mid-winter and the heather honey later, when breeding gets underway. Better for the bees, and for the hive hygiene.

(c) If neither (a) nor (b) apply, your brown smears may indicate nosema or some disease problem, but just could be caused by food problems, unripened stores, for example.

Bumblebees

Usually in April there are some very large queen bumbles, having recently emerged from winter hibernation, which are looking for nest sites and can often be seen flying under a hive, sometimes even going in at the entrance. I have never seen one attacked by bees, but they usually come out and fly away. Later in summer, the smaller bumble workers will sometimes go in, perhaps attracted by the smell of nectar, but do no harm and come out unharmed. It almost seems that honeybees know them to be friendly cousins, yet usher them out firmly.

Acarine Testing

While seated by a hive, it is quite simple to examine freshly dead bees for acarine, impaled on a cork by two pins held close together, as shown in *Figure 1*. With a watchmaker's eyeglass pressed into one eye, push off the bee's head with the small blade of a penknife and scrape of the collar to reveal the large trachea (normally white and semi-transparent with coiled cartilage reinforcement, something like a vacuum cleaner tube). Any discolouration will show up clearly, possibly on one side only, possibly on both sides. Finding one bee clear tells you nothing - you should test several to get

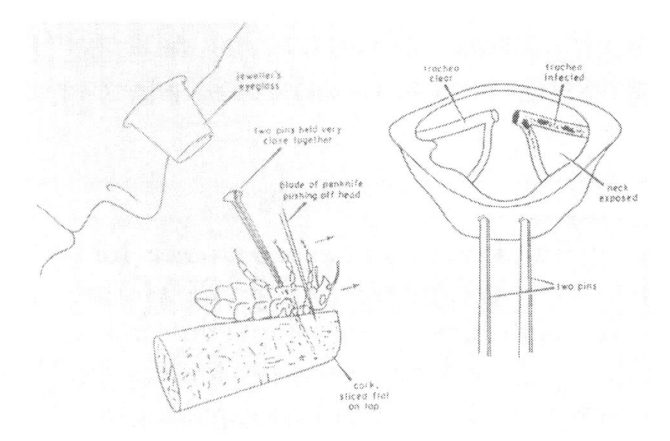

FIGURE 1. Hive-Side Acarine Test.

some indication.

Obviously one would choose to test a hive having a number of dead or dying bees on the alighting board, with perhaps others crawling on the ground as if trying to get away from the hive.

Drones Flying Freely

This is normal during the warmer hours of the day from late April to the end of July, and does not necessarily mean that the stock will swarm. No colony is happy without some drones in summer, and even young queens will produce some. Older queens (3 or 4 years old) will normally produce far more drones than younger queens, and are more likely to swarm.

Honey Flow

When the air is full of flying bees, in a hurry to get in or out, sometimes flopping down on the ground and panting to get their breath back, when eager bees are climbing rapidly part of the way up the hive front, pausing for a moment to comb their antennae clear of pollen before flying away; when the air is heavy with the srnell of nectar - then you have a 'honey flow'. Bees are collecting far more than is needed for daily use and the hive will be increasing in weight by 2-10 lb. a day.

Swarming Bees

The possibility of swarming can be a problem for beekeepers who cannot keep up with regular inspections in early summer, so look out for the signs. Bees loafing at the entrance, nibbling aimlessly at the woodwork, running a few steps and looking up and down; above all, other bees exploring old boxes in a shed or garage. All this suggests the strong likelihood of a swarm in the next few days (or hours!) If you have a glass crown board, lift off the roof and you may see rows of worker bees lined up on the sides of central top bars like soldiers on parade - 'swarm parade' - waiting for the word 'Go'.

Swarm-watching

Not strictly hive-watching, but at least bee-watching. If you ever have a swarm conveniently placed in the garden, make yourself comfortable and settle down to watch them for an hour. You may see the queen walking on the surface for a few seconds before diving back into the cluster, and you may be able to pop her into a queen cage and hang it alongside. You may even see a second queen, perhaps a virgin out with her mother, or two or three virgins with a cast. You will almost certainly see scout bees performing the 'von Frisch dances'. Very often there may be two different dances on opposite sides of the cluster, indicating two different sites chosen by scouts. After a time (which could be an hour or a couple of days), the debate is resolved and the same dance will be seen in perhaps three different places on the cluster. The bees have been debating the relative merits and have arrived at a consensus: like, for example, a hollow tree 1200 yards north-west, rather than a chimney 900 yards southeast. Any time after this they will take off and fly in the right direction, led by scout bees and finding other scout bees at the chosen site, fanning with Nasanov glands exposed to welcome them in. Using the graph (Figure 2) you may be able to read the dance and know beforehand where the bees have decided to go.

If the suggested site is very near to the clustered swarm (less than 100 yards), scout bees will perform rapid circular dances with no directional information, so that the bees will be led directly from swarm to site by scouts, with a Nasonov beacon to guide them in. I have noted this only once, as it is more usual for bees to go directly from their hive to the new site in such cases: for example from a hive in my garden to a pile of loosely stacked, empty supers 30 yards away (noted a dozen times at least).

More usually swarms form clusters first within a few yards of their parent stocks, and choose a new home at least several hundred yards away, but remember always that bees do nothing invariably! Scout bees which have chosen a new home communicate by 'figure of eight' or 'wagtail' dances, in which the straight central line makes the same angle with vertical as the flight line must make with the sun's direction. Even if the sun is not directly visible, its direction will be perceived by the bees either from a small patch of blue sky anywhere or by sensing polarised light coming through thin cloud cover. In Figure 3, the dance angle is about 30' to the right of vertical (at one o'clock), so the flight direction must be at 30' clockwise from the line or plane joining swarm to sun. Distance to be travelled is indicated by the frequency of straight runs - rapid for short and slower for more remote sites. In practice it is easier to count the number of straight runs per quarter minute, when the distance can be read off from the graph (Figure 2). This is necessarily an abbreviated account. Read from von Frisch's book, "The Dancing Bees", for a full treatment.

White Grubs Thrown Out

Three different points can be made here:

(i) If the grubs are long and thin (possibly like small white caterpillars), then it shows that wax moths are present, but are at least being dealt with by the bees. It suggests that some combs are perhaps old, black and should be replaced with frames of wax foundation as on occasion offers. Perhaps later on the colony could be transferred to a clean brood box (either new or an old one well scorched inside with a blow lamp).

(ii) Large, white grubs suggest that drone larvae are being ejected, possibly because food is scarce, so feed sugar syrup. Very occasionally it can mean that a young mated queen is now laying well and that no more drones are thought to be needed this season.

(iii) If the ejected larvae or pupae are smaller, i.e. worker brood, this is a sign of starvation and emergency feeding is called for. I would recommend the use of a plastic squeeze bottle (ex washing-up liquid) to squirt warm syrup directly

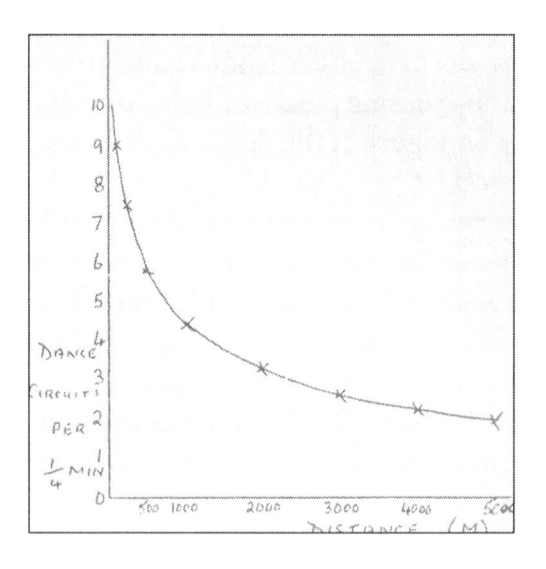

Figure 2. Von Frisch Dance.

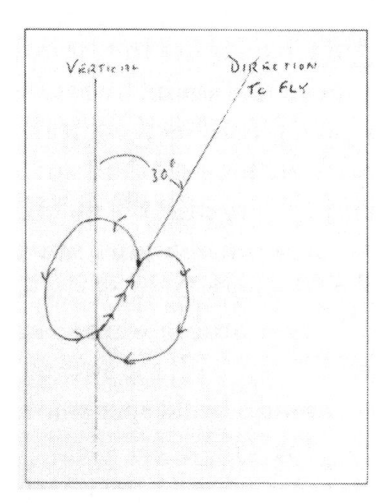

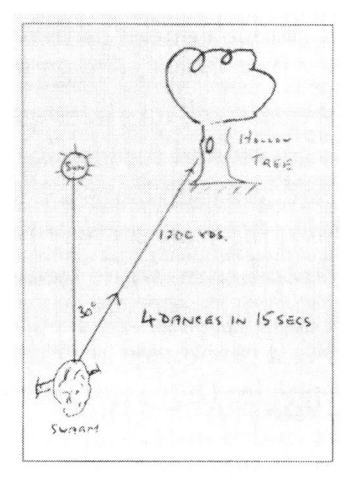

FIGURE 3. (i) and (ii) Direction and Dis-tance of Swarm to New Home.

into one or two combs, and follow this with half a gallon in a normal feeder. Do not worry about sugar syrup getting into honey: there won't be any if worker larvae are being thrown out, and dead bees don't store honey anyway. Feeding bees in June may be unusual, but in the fickle British climate it sometimes has to be done, and the bees will take the food straight down into the brood chamber, whether a super is on or not.

Sounds

All the books mention 'piping' queens: that very high-pitched note clearly audible from six feet away. This sound is made when one or more virgin queens are at large in the hive. It may mean that they contemplate swarming (but have not done so yet, perhaps in poor weather), or more likely that the colony swarmed the day before. A totally different sound is 'quarking', a deep, hoarse sound repeated six or seven times, Ark, Ark, Ark', etc. as described by Dr. Bevan in his 1837 book (but not mentioned by modern authors). This sound is made by young queens still imprisoned in their cells, and can attract a queen or free virgin plus worker to tear down the cell and destroy her. I have taped this sound by resting a ripe queen cell on a microphone and recording it while she is sawing around the capping before pushing off the trapdoor (heard as a loud bang as it swung open against the mike). In a hive, the workers may be waxing up the virgin's saw cut to keep her in, while feeding her through 2mm of the saw cut kept open for the purpose. They probably want to keep her as a new queen after the one has left with the swarm, or even to lead out a cast later on. Should a beekeeper smoke a hive in this interesting condition, the worker bees leave the queen cell(s) and within two to three minutes the queen(s) escape(s). I still remember one afternoon 20 years ago, at an out-apiary when I had several virgins free in a hive, even two crawling up my left arm!

In a quiet time of day (late evening or early morning) one may tap a brood box twice with bare knuckles and hear the response. A happy queen-right colony will give a sudden 'Hiss' that dies down as quickly as it started. On the other hand, a queenless colony will give a long, moaning roar sounding a bit like the three witches of Macbeth or a cow a mile away who has lost her calf. Not part of this story, but when opening up a hive, if one frame is accidentally jerked as propolis unsticks, the operator can quickly say "This hive is queenright", while onlookers say, "How can you know that?"

Propolis Collecting

This is not often described in modem books, but Francois Huber in 1798 described how bees pull the sticky material from fat, red poplar buds and pack it in their pollen baskets in small, glistening globules "the colour and lustre of a garnet". This description has never been bettered, as you will immediately agree when you see a bee waddling into your hive with bejewelled rear legs. Remember that those were the words of a blind beekeeper seeing through the eyes of his assistant, Burnens, and if you haven't yet seen it yourself, find an old QX or crown board with propolis deposits and stand it in the sun on a hot day in July; before long several bees will be investigating, and you can watch with amusement how hard they pull to detach a piece, with legs pressed down to get a good grip, and then sometimes fall over backwards with complete loss of dignity as it suddenly gives way! Now watch very closely to see how the propolis is transferred from mandibles to rear legs. A chewed-off lump may not resemble Huber's garnet, but you will see that one day, fresh from original sources, if you persist. Inside the hive the house bees may just pile it up for future use or work on it immediately, mixing it with saliva to make a varnish with antiseptic qualities, or in several other ways.

BKQ 38
Summer 1994

Wasps Trying to Enter

- a hive (usually in July / September) but being intercepted and thrown out, each one by two or three bees. Perhaps some dead wasps can be seen on the ground below in front of a hive. This is a sign of good hive morale, of a strong hive successfully defending itself. You can help by stubbing a wasp on the alighting board with your finger, and leaving the bees to complete the job as part of their colony defence training.

Wasp nests should be located and destroyed in July when their usefulness is finished.

Wasps Entering Freely

- with occasional bees half-heartedly challenging them. This

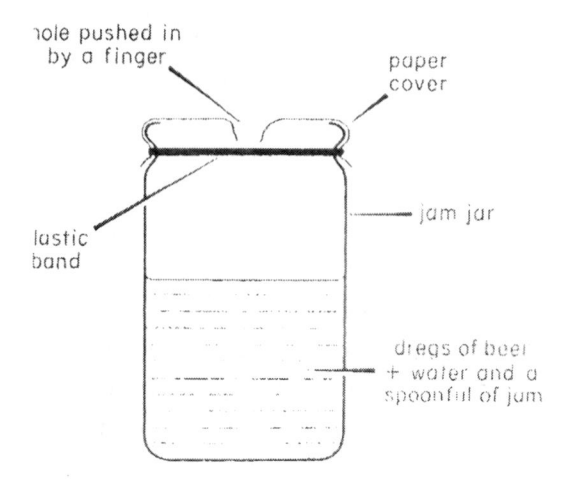

A simple wasp trap. Wasps can be a nuisance in August and September. One or two traps like this will divert them from the hives and reduce their numbers. Use no sugar or honey, though, or bees will also be trapped.

indicates a weak colony liable to be robbed out if no action is taken. Restrict the entrance to half an inch and, if necessary, move the hive to a site at least two miles away. Hive entrances should routinely be restricted by the end of July to the orthodox narrow part of the entrance block.

A hunt around the local area may disclose an active wasp nest too close for comfort and needing to be destroyed. I do realise that wasps are also useful insects for most of the summer, and I have seen them taking grubs and small caterpillars from raspberry canes and other produce. By the end of July they have normally finished most of their breeding and are searching for carbohydrate food only like honey and plum juice; my sympathy for wasps then evaporates rapidly.

Ants

One may notice ants at the entrances with guard bees fanning vigorously to try and blow them away. Ants are too small to offer a target for stings and too hard to bite; bees find them very irritating, especially when two or three ants decide to cling to a bee's legs. They are not usually serious enemies. One should avoid apiary sites close to existing ant hills. Years ago at Lustleigh I had to abandon a small apiary site on the edge of a wood after one hive was taken over and two other hives seriously distracted by large redwood ants. In Central Africa ants of one kind or another were much more dangerous.

Drones Being Ejected by the Bees

Usually this happens in late summer/ early autumn (end of July onwards), when swarming or supersedure is no longer expected. I have seen it happen rather earlier than this and on checking up have found a young queen arising from supersedure at the end of June. Obviously this is related to the possible needs of the colony, heartless though it may seem to sympathetic humans.

Very Small Drones

When seen flying or walking on the alighting board and a shortage of worker bees are an indication that you have a queenless colony with laying workers, or a drone laying queen; in either case, the colony will not survive the winter and action must be taken. On opening up the hive this will be confirmed by the presence of high- domed sealed worker cells scattered around the brood area. The presence of several eggs in one cell characteristically on the sides rather than on the base, will confirm the presence of laying workers. Such a stock may be saved by throwing in a swarm if one is available when the swarm will take over by sheer pressure of numbers and energy. It is usually hopeless to attempt to re-queen by any normal methods and often the best answer is to move the hive to one side, open up and shake the bees off all combs. Many non-laying workers will be accepted by other stocks and revert to normal. The brood box plus combs (no bees) may be placed on another hive to be cleaned up. A well-known French beekeeper (M. Jean Scrive) tells me that he has successfully re-queened a worker brood colony by painting a new queen with royal jelly and just dropping her in.

Robbing Bees

These can resemble bees in a honey flow, but a closer look will reveal many differences. Their stomachs will be empty so no heavy abdomens. They usually hover for a moment with hind legs hanging down to have a good look before deciding to enter. Their bodies can be shiny from the attentions of nibbling guard bees. they can look like chronic bee paralysis virus sufferers. Above all they will never have pollen loads on their legs: robbers take food out of a hive, not into it. If you look at other hive entrances in the same apiary you will notice clusters of guard bees everywhere saying, "You won't get in here". You may sometimes find that just one hive a few yards away is doing the robbing with a shuttle service flying between the two.

What can one do about all this? First aid can be by narrowing the entrance or even stuffing grass in for an hour or two, but robbers have long memories and will come back. If it is a nucleus being robbed, then stuff the entrance with grass and at dusk, when flying has stopped move the nucleus to the other end of the garden, or preferably to a site at least 200 yards away. The most interesting solution is to interchange hives i.e. put the robbed hive on the site of the robbers and vice versa then sit back and watch the robbers taking honey back to the hive they have been robbing! After a time the bees get thoroughly bemused, badgered and bewildered and give up robbing. One feels like a judge who has made a court order for restitution. This involves too much work to be a practical solutions but remains a fascinating experiment in bee behaviour, contriving a situation which would never occur naturally. I have done this two or three times over the last 30 years (and wasted a couple of hours enjoying the spectacle).

Death of Flying Bees (Summer)

In a suburban garden one may have bee losses arising from an unknown amateur gardener half a mile away who sprays his two apple trees with more enthusiasm than knowledge: this can result in several hundred dead bees on the ground and alighting board with perhaps only one hive affected where the previous day no such problem existed. One is unlikely to discover the source and not much can be done about it, except to clean up and hope that it won't happen again and to pass the word around in the hope that the culprit will come to know the damage he caused. On a larger scale there are spray warning organisations giving advance warning of crop-spraying especially where numbers of hives have been moved to pollination sites. The amateur gardener will not usually be involved in this and it remains a hazard, fortunately only a minor one.

Cluster of Bees on Hive Front

A broad cluster over the front of a hive, with perhaps a 'beard' of bees hanging down from the entrances may indicate just overcrowding inside and the need for another super so try this first. Sometimes it means something quite different. I have known a 'beard' of bees to persist, even in the rain, after I had taken the trouble to give them empty supers and even an eke under the brood box for spare clustering space. Such an external cluster can sometimes block most of the entrance as well as covering a third of the hive front, with stubbornly immobile bees wet in the driving rain. Why? It does not necessarily indicate that a swarm is imminent but a check for swarm cells would be a reasonable precaution. In my experience these bees are not wax-workers except very rarely when they are a swarm from another hive usually clustered on the rear of the observed hive rather than the front.

If they already have plenty of room inside it suggests a surplus of nurse bees without enough work to do, the queen having reduced her laying in July. I have even heard it suggested that nurse bees with swollen brood food glands feel the same discomfort as unmilked cows with swollen udders and seek relief in the cool rain. I really must remember to open up a dozen such bees next time under a dissecting microscope and check for turgid hypopharyngeal gland buds. Your observations and ideas on this subject would be welcome. What should one do about it? Well, often this condition declines after a few days, especially in a good honey flow, even if nothing be done. Alternatively the hive could be slid three feet sideways on the stand and another weaker stock put in its places to collect the flying bees and reduce the population.

Gaps Between Supers

With old equipment sometimes bees may be seen using an unofficial entrance high up on the hive between two supers; this happens more often at a corner, perhaps where a hive tool has been used too energetically in the past, or where the timber has shrunk at a joint. The bees appreciate a 'short cut' to where they have to store the honey and no harm comes in high summer. But remember in autumn, when putting on clearer boards, that this is where the bees will get in to take honey down into the brood chamber, and in 48 hours the super may be half-emptied, with torn down cells, and hundreds of bees from the hive next door joining in. They will be so excited that much of the stolen honey will not go to increase winter reserves but just be used up with no gain in stores.

Hefting Hives

Looking and listening will never tell you the weight of a hive but often one needs to have some idea of how much food has been used, without opening up. If all the hives are of the same type, made with the same thickness of the same type of

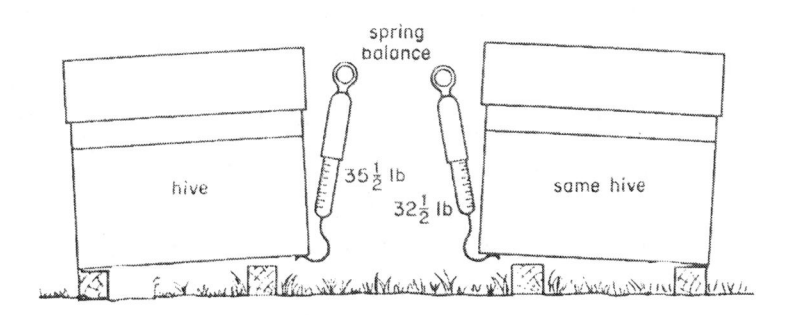

Weighing Hives. Total Weight = 68ib

wood, just lifting one end off the ground by hand can give a general impression of lightness or heaviness, for the time of year. Unfortunately this is seldom the case, but with a spring balance graduated in 1/4 lb. or even 1/2 lb. notches, a sufficiently accurate assessment can quickly be made. Just slip the hook in turn on each of two opposite sides, and note the weight as the end of the hive just lifts up; add the two weights to obtain the total hive weight, and compare this with your record of the weight of that hive last autumn. (See diagram). Obviously the far end of the hive must rest on a runner, not balance at a point even a short distance in.

It is quite surprising on occasion to find just one hive, out of perhaps ten in an apiary, much lighter than the others, although all had been fed in September. Finding this out in February enables a beekeeper to feed where necessary, to keep the stock alive.

BKQ 39
Autumn 1994

From December to February it would be usual to pay just an occasional visit to check that all's well after a gale, that no roofs have been blown off or tree boughs fallen across, or farm animals have pushed over hives. Bullocks mean no harm but if they get into an apiary, they may use a hive as a scratching post and displace boxes, giving access to mice or robbers, I have twice found a displaced roof, with yellowing grass under telling me it happened at least two weeks before. What to do? Well, I once found seven brood frames in the open air stuck together with propolis, with clustered bees. They were put back in the hive, fed a gallon of warm sugar syrup (mid-January and frosty) and by May were as strong as any other colony. I am not recommending that syrup should be fed in January, but without food they would have died anyway. It just shows how tough bees can be.

Another point to watch on occasional visits to out-apiaries is the ground in front of a hive entrance. Any dead bees? A few would be normal, several cupsful would indicate trouble. None at all would be a good sign, or else that efficient birds or mice had cleared the area. Look out for wax debris too (mentioned already).

Smells at Feeder Hole (Winter)

Lift off the roof very carefully, without the slightest sound or knock and apply your nose to the feeder hole; be careful not to kick the hive as you do so. What smell do you notice?

(i) A cold, damp, musty, churchyard, abandoned building smell: The hive is dead and should be opened up to see why. Absence of stored honey, with dead bees, heads in cells means starvation, often with ample stores of pollen. The same with no bees on the combs or floor and combs perhaps showing signs of robbing, with ragged edges, could mean queen failure in winter and wise bees laden with food being accepted next door. If any sealed brood is still present, this must be opened up and checked. Finally, block up the hive entrance.

(ii) A homely smell, as of newly baked bread, very young freshly washed and warm human babies, means that all is very well, with queen laying and open brood present. Replace roof and walk quietly away.

(iii) A strong smell of mice (acetamide to chemists) or of ammonia, or urine. There are mice in the hive. Open up on the first mild day in February to remove frame and mouse nest, replace with two heavy food combs and change the floor.

(iv) A sour, offensive smell will suggest a brood disease: to be opened up and more accurately diagnosed on the next mild day.

(v) Fresh nectar - even small traces can be smelt - probably not before early March, but always welcome to note as a good sign.

Apparently Lifeless Bees on the Ground

Often on a cold day in February or March one may see bees on the garden path or just in front of a hive, perhaps with a load of pollen, apparently lifeless. Try picking one up, cupping it in one hand covered with the other hand and breathing warm air on to the bee. After four or five breaths it will probably stir and then try to walk. Carry on for another minute and then launch it into the air, when it will complete the journey home, refreshed by your 'kiss of life'. It may have been a bee almost ready to die of old age, but it could be a water-carrying bee with cold water chilling its honey stomach, so that its muscles are too cold to develop enough power for take-off. Since it has only the 3% sugar content of its haemolymph for fuels it will have had a limited flight range. I think of this as resembling a human out shopping and hoping to get home on the petrol in the carburettor. The answer to the water problem, needed by bees in spring to dilute the concentrated stores for larval feeding, is to provide a source of water a few yards from your hive. An old trough of water with pieces of bark or peat placed in a sunny corner will save many bee lives in Spring.

Slugs

Sometimes large slugs are found in hives and bees cannot deal with them. No great harm is caused, and I usually flick them out with disgust. They may feed partly on pollen and hive debris, never on combs, in my experience. They are less likely to gain entry if hives are further off the ground: on hive rails, for example.

Woodpeckers

- are normally only a nuisance at out-apiaries and in unusually cold winters (like 1963 for example). They usually go for the sides of a hive, especially if 'D-shaped' cut out handholds are there. In nature over tens of thousands of years, they must often have come across wild bee nests in trees, while searching for grubs. When desperate for food they obviously associate hives with bee nests in trees and peck holes to get at the stores.

Prevention is simple, either by tacking half inch mesh galvanised chicken wire netting around the hive, or just draping a plastic farm sack so as to hang down from the roof.

Temperature and Flight

It is often said that dark bees work at lower temperatures than yellow bees, for the quasi-scientific reason that they absorb solar radiation better. A half-truth here, for when the sun is not shining, dark bodies lose more heat by radiation than lighter coloured bodies. From personal observations sometimes lighter bees will work at lower temperatures. They all seem very conscious of air temperature and if after a cold day there is a sharp increase in temperature even as late as 3.00pm, I have seen four different hives all start up flying within a minute or two of each other. There is plenty of scope here for hive-side research.

Instinct or Intelligence?

I quote this story last of all, but it was actually my very first hive-side observations which in 1953 changed the whole course of my life. A swarm of bees came into my garage in Zomba Road, Lusaka, Zambia and occupied an empty wooden packing case. That same evening my cook Mafafa Phiri and I carried out the case into the garden and put it on four bricks between two papaw trees, with a sheet of corrugated iron over as a roof plus more bricks to keep it firmly in place. I was busy for the next few weeks and used to have an occasional look to see if they were active, but saw few bees flying. The rains had ended early that year and there were plenty of flowers about. Then one day early in April I noticed that bees were dragging fibrous material out of the gap between case and lid, which they were using as an entrance. A closer look showed that this material like finely spun yarn or hair, was really a mass of shredded brown paper. I realised then that the packing case, ex Baird and Tatlock scientific equipment, had been left stuffed with the Kraft tarred packing paper used to wrap individual items of laboratory apparatus, as well as to line the case. This specialised packing material consisted of a layer of tarred paper sandwiched between two layers of glossy brown paper to form an extremely tough and damp-proof protective covering. Obviously, I thought, the bees were expanding and needing more space so were chewing away the paper.

My admiration for their industry grew and I made enquiries of my staff; one lady had a husband in the Forestry Department who had kept bees years before in England, and he lent me the first book on beekeeping that I ever saw, "Honey Bees and their Management" by Stanley B. Whitehead. My previous knowledge of bees was limited to three items: that they made honey, had stings and sometimes swarmed. This book opened my eyes and I made exhaustive enquiries to see if there was a beekeeper around, but alas in a population of over 60,000 no one kept bees, though many warned me of their bad temper and dangerous stings. I wrote by airmail to an address quoted in the book, Robert Lee & Co. of Uxbridge, with a £25 cheque to open an account and asked for a catalogue. Then one day in May I first noticed that among the daily little pile of finely shredded paper there were one or two curved strips about 5mm wide and 4 to 6cm long, clearly showing by bite marks that they had been cut out by the bees. Within a week the output of shredded paper ceased completely and bees were to be seen dragging out the paper strips, even flying away with them, sometimes two bees together, and drop-ping them well away from the packing case. I was most impressed. Surely this was a remarkable example of applied intelligence, firstly to discover a more effective way of solving a problem, then to have all the work force changing to an entirely different and much more efficient process so quickly, with no opposition from older bees or bee trade unions!

I consulted biologist colleagues on the staff, but they all told me that insects had instincts but no intelligence as we understood the term. I argued that bees could never have come across modern, sophisticated tar-paper in any hollow tree, and that if humans had reacted to a problem in this way it would have been described as an example of intelligence, so why discriminate against insects? For the record, I glued specimens of both shredded paper and the curved strips into my bee diary, and a photograph of this is shown here. From this moment I became fascinated, even obsessed with bees and made my own hives, imported a veil, smoker, a single sheet of zinc queen excluder, but mostly improvised. I made many mistakes but I learned from them, in total isolation. Then a friend from Natal gave me an old copy of Roots' "ABC and XYZ of Bee Culture" dated 1920, having 850 densely packed pages. I read this book for an hour or two a day for a year or more and have it still. My African bees had never read this book and didn't always behave as they should, but it helped me enormously. I worked out my own system of management and within two years was taking a crop of over

SECTION 1. LOOK, LISTEN & LEARN

Shredded Fibres and Cut-out Segments

100lb. of honey per hive a year and selling it at a good prices with no competition at all (except from imported honey). I never owned an extractor, using only home-made hives (wood from old packing cases), made 12 queen excluders from one by cutting up squares and letting then into square plywood boards, and devised various items of equipment and techniques, some of which I still use here in England. The bees were indeed very bad-tempered, but knowing no better, I thought all bees were like this and worked out techniques of coping with them. On one occasion my pen of ten laying hens were stung to death, sometimes we were imprisoned in our house from lunchtime to dusk and so on, but I gradually learned ways of handling these intelligent creatures and during my last few years in Central Africa never had less than six hives in my garden, in the middle of towns like Lusaka and Ndola.

All this started from my first 'observation at the hive-side'!

SECTION 2.
ECONOMICAL BEEKEEPING

These accounts of economical beekeeping are based on the experience of over 40 years with bees; in Central Africa with no access to bee appliance dealers and later in Devon where the habit of 'make it yourself' persisted.

BKQ 41
Spring 1995

There are three main aspects to this topic:-
(i) the construction of various items of equipment which are more suitable for their purpose than anything on the market and also much more economical,
(ii) the production and marketing of products other than jars of honey,
(iii) the development of techniques which are more productive, less physically demanding and also better for the bees than those currently in use.
This article deals with some examples from the first category.

Equipment: Queen Excluders

For the first twelve years of my beekeeping I was in Central Africa, with no access to shops selling bee appliances. I did however obtain a second-hand zinc queen excluder from a friend in South Africa, and from it I cut out sixteen squares which I set into sheets of framed plywood and used very successfully on my hives. Since my return to the United Kingdom over thirty years ago, I have followed the same pattern, except that I have used longer strips, as shown in Figure 1. That shown was cut from a damaged W.B.C. Waldron excluder, but three inch wide strips cut from old zinc excluders are even easier to use. I prefer the short slot variety, as it is more rigid.

The Waldron excluders are cut so as to leave wire ends projecting at each end, also cross-pieces at each side. These can be pressed into the central section of 3 ply wood and the edges reinforced with Cascamite to give extra strength. The bees usually propolise the points of contact between metal and wood, making them stronger still.

These have worked extremely well, even when in an exceptional year the bees have needed six supers above the Queen Excluder. One bonus, as against full-size zinc sheets, is that supers separate more easily from the flat and rigid wooden surface of my pattern, with less sticking when lifting off.

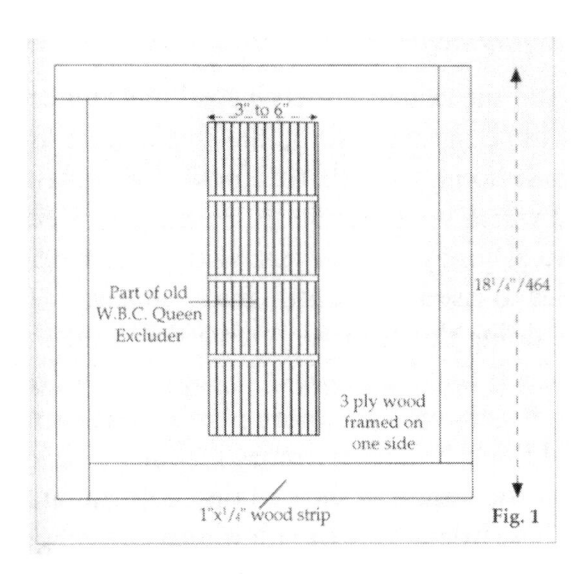

Fig 1. Economical queen excluder

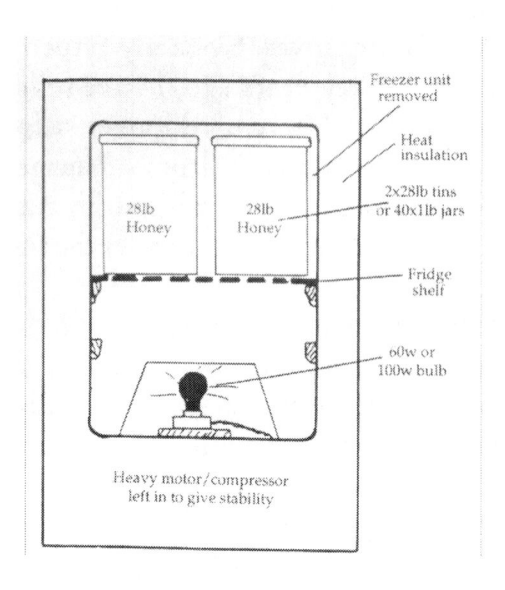

Fig 2. Home-made honey warming cabinet. Adapted from old household fridge (door not shown).

Equipment: Honey Liquefying Cabinet

I have found the honey-liquefying cabinet (Figure 2) made from an old refrigerator, to be the single most useful item, as I stockpile honey in 28lb. tins (now also in plastic bakers' pails) and warm up about 56 lb. a time for filtering into a stainless steel honey tank. After twenty-four hours to allow the bubbles to rise I then bottle into 1 lb. and 1/2 lb. jars. I have been doing this for twenty-five years; the only modification now is that I use bricks to support the fridge shelf. Modern refrigerators have more plastic and less metal than my first old one, and so are more inclined to buckle internally under the weight of the two tins, deforming the plastic interior. These bricks have the additional advantage of acting like a 'night storage' heater, absorbing heat and then continuing the warming-up process after the bulb is switched off. I also have the bulb resting in the cavity of a brick, having

SECTION 2. ECONOMICAL BEEKEEPING 43

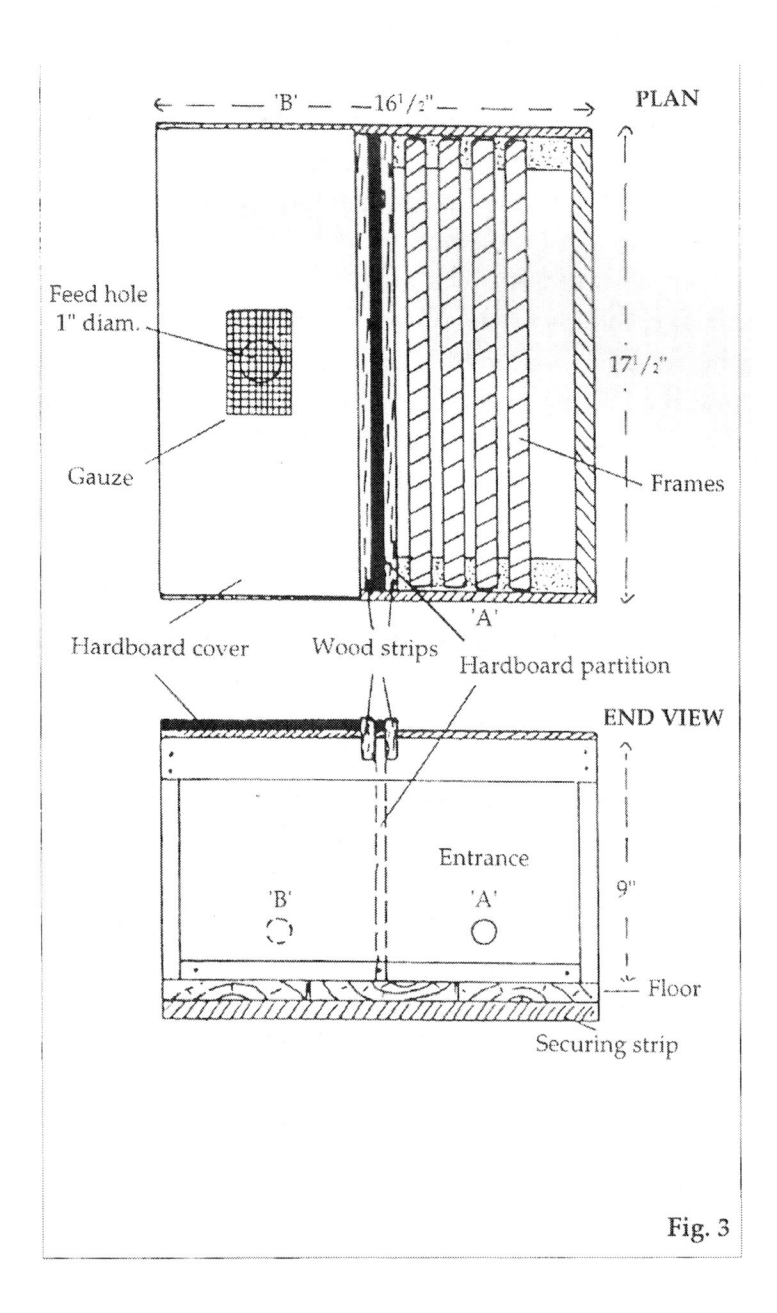

Double Nucleus from old W.B.C. box

been somewhat surprised to find it slowly burning a hole in a wooden support. When I first adapted a fridge for this use, I found 'hot spots' vertically above the bulb, but I completely overcame this problem by bending some thin sheet aluminium into the form of a bridge, to disperse the heat from directly over the bulb. I have never needed to buy an old fridge: in fact I was once offered £5 to take one away, but declined with thanks. The more efficient the insulation of the fridge, the lower the power of the bulb needed to bring the honey up to a temperature of about 120°-130°F in 24-36 hours. If a 100 watt bulb is used it is best to switch off for a few hours after a twelve hour stretch. Obviously it can also be used to liquefy granulated or partly set honey in jars.

Equipment: Double Nucleus From Old WBC Brood Box

The double nucleus boxes made from old W.B.C. brood chambers have also seen good service for many years. There are still so many old W.B.C. hives about that one can often obtain them for nothing, or at least for a very low price. Bees over-winter well in them, usually clustering towards the central partition so that each nuc helps to keep the other warm (Figure 3).

Feeding is via pint-sized round feeders containing syrup, or plastic containers (ex margarine) of bakers' fondant with five 1cm. diameter round holes in the lid (punched with an old fountain pen top). A Smith or W.B.C. super encloses the feeders, with a National roof above.

Equipment: Solar Wax Extractor

In 1970 I first adapted an old W.B.C. outer lift into a solar wax extractor. The obvious initial advantages are:

(a) the inverted pyramid shape which allows a maximum entry of solar radiation; (b) the built-in plinth (made to fit another outer lift when on a hive) which provides a ready-made seating for a rigid sheet of clear plastic. Orthodox solar

SECTION 2. ECONOMICAL BEEKEEPING

"Greenhouse Effect"
Solar Wax Extractor

Painted dull black inside glossy white outside
20" sq. thick plastic
Old W.B.C. lift
Plywood
Insulation
Old floor boards
Metal tray
Course gauze filter
Tin for wax not shown

Fig. 4

Rotating stand
Pipe sunk in earth

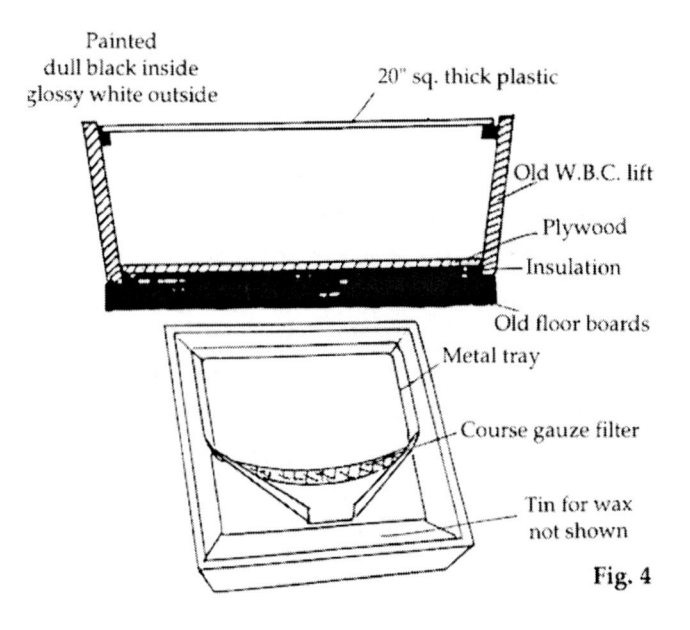

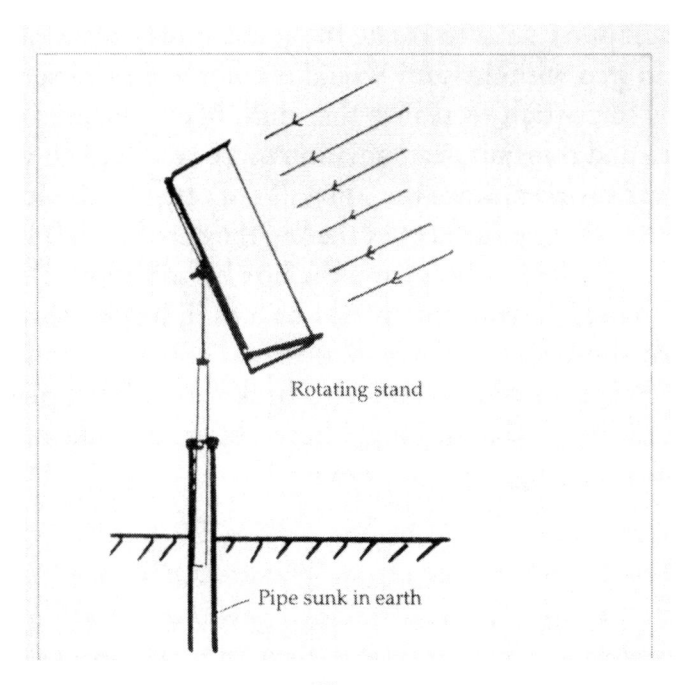

Figure 5

extractors use double glazing, but I soon discovered that even a 1/8" thick transparent plastic sheet was just as efficient; also it was less liable to break when my large Bramley apples fell on it! Glass is a relatively poor insulator. If you doubt this, place your hand on a sheet of glass; it will feel cold, as metal would. Do the same with a sheet of plastic at the same temperature and it will feel warm. Glass conducts heat away more rapidly; the plastic much more slowly. It is important to fix a thick wooden floor, with a sheet of thin polystyrene covered with plywood, otherwise heat will be rapidly lost. I have also found it advantageous to support the metal tray so that it is fairly close to the plastic sheet, but not so close as to preclude reasonable room for wax combs or fragments to be melted. Some years ago I added the refinement of a rotating stand (Figure 5), so that the extractor could, at the touch of a finger, be made to face the sun, but also with a facility for up-and- down rotation to catch the solar rays at right angles. I am waiting for a real inventor to add a small motor coupled to a computer so as to rotate the stand slowly from east to west in step with the sun! Would a sunflower plant serve?

It is important to realise that dull, black surfaces both absorb and transmit heat radiation more readily than white or silver surfaces. Since the sun will not often be shining on the outer sloping surfaces of the lift, they are best left white to minimise loss of heat from the box by radiation. On the other hand, the interior should be a dull, black colour to absorb short-wave solar radiation and convert it to long-wave radiation which cannot go back through the plastic. On a really hot day in June, I have obtained a cake of fine yellow wax weighing 3 lb or more.

> *The surface of the sun at a temperature of about 6000 de-grees Celsius emits mostly short-wave radiation, whereas the black interior of the solar box will probably be at a temperature of only 100 - 150 degrees Celsius producing long-wave radiation which is trapped inside (greenhouse effect).*

SECTION 2. ECONOMICAL BEEKEEPING

Wax from Old Black Combs

Experience shows that old brood combs which are heavy and black will not yield any worthwhile amount of wax in a solar extractor. This is because old brood cocoons and faecal pellets in the cells soak up the wax. To obtain a worthwhile yield, use only combs with at least a trace of yellow colour in the solar. Black brood combs should be broken up and soaked in rain water (or tap water plus half a cup of vinegar per two gallons) before boiling as shown in Fig. 6. An occasional press on the sack of combs while the water is boiling will help. Overnight the wax will set into a disc, easily released by running a knife round it; the base of this disc will be black and much of this can be scraped off. The disc can then be re-cycled through the solar and partly bleached by the sun. When dried, the old combs can be used as firelighters or added to the compost heap.

BKQ 42
Summer 1995

Warming Honey Supers

It is often said that honey supers straight from the hive can be efficiently extracted, but some years ago I made comparative trials, using an old-fashioned steelyard. hanging from a beam in my honey shed and a length of thin rope, wrapped around the supers. Accuracy was only to the nearest 1/4 lb. Having recorded the full weight and taken the temperature of the combs by pushing a thermometer into the cells at an oblique angle, the combs were spun out and replaced for a second weight to be taken (extracted weight). The supers of wet combs were then left on a hive in the garden for 24 hours, until they were licked dry and then weighed again (dry weight). This was repeated four times with different supers, so that an average result could be calculated.

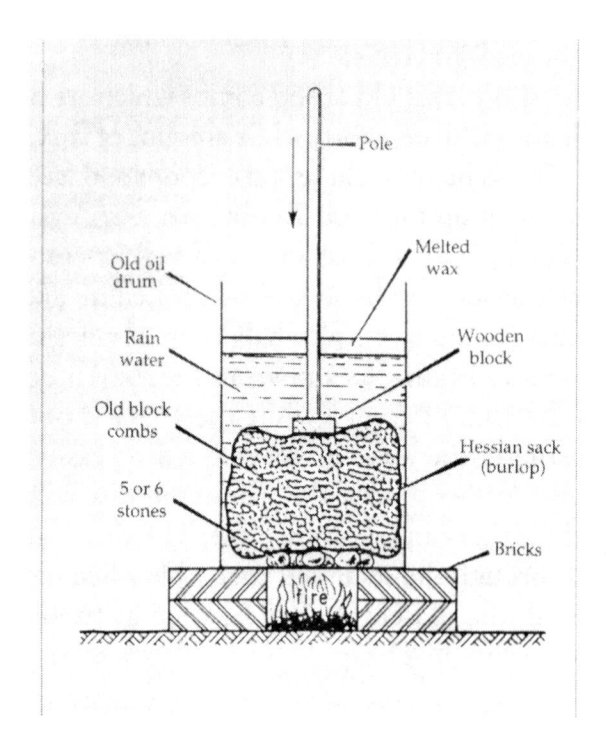

FIGURE 6. Wax from old, black combs

As can be seen in the table below, the weights of honey extracted and left behind in the cells were easily obtained by subtraction. The results for four different supers agreed very closely, giving an average of 3 lbs. honey left behind on wet combs for every 25 lbs. extracted, i.e.12% when working at a temperature of 60 degrees F / 16 degrees C, with supers straight from hives in a nearby apiary.

When I repeated the experiment with four other supers at a temperature range of 70 - 85 degrees F (24 - 30 degrees C), the honey left behind was reduced to an average of just under 1 lb per 25 lbs extracted. This effect is very well understood by professional beekeepers, who find it worthwhile to build a 'hot room' to bring honey supers up to a suitable temperature. Even in subtropical Queensland, Norman Rice

SECTION 2. ECONOMICAL BEEKEEPING

Honey Left Behind in Extracted Supers

Full Weight	Extracted Weight	Dry Weight	Honey Extracted	Honey Left
$39^1/_2$ lb	$13^1/_2$ lb	$10^1/_4$ lb	26 lb	$3^1/_4$ lb
37 lb	$12^1/_4$ lb	$9^1/_2$ lb	$24^3/_4$ lb	$2^3/_4$ lb
$35^1/_2$ lb	12 lb	$9^3/_4$ lb	$23^1/_2$ lb	$2^1/_4$ lb
38 lb	$12^1/_2$ lb	$9^1/_4$ lb	$25^1/_2$ lb	$3^1/_4$ lb

At 60°F/16°C, 3lb. honey left for every 25lb. taken

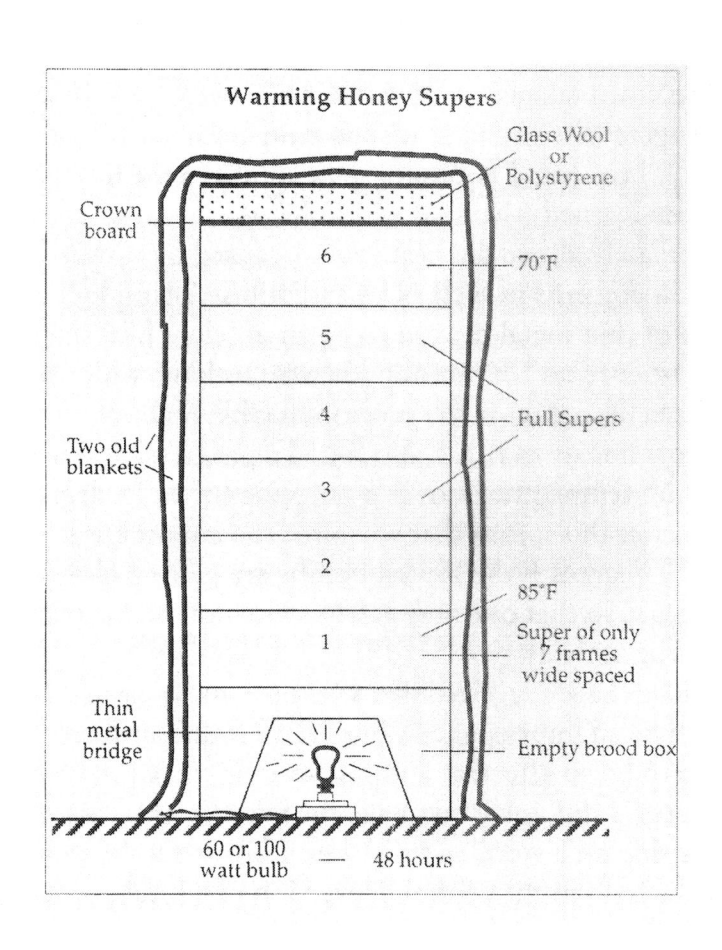

Warming Honey Supers

Glass Wool or Polystyrene
Crown board
70°F
Full Supers
Two old blankets
85°F
Super of only 7 frames wide spaced
Thin metal bridge
Empty brood box
60 or 100 watt bulb — 48 hours

found that the extra honey extracted more than paid for the cost of warming it up. I have heard it argued that the bees should have the honey they could lick off, for winter store, but I found that the excitement caused and the extra activity generated had the result of reducing the gain in weight to negligible proportions. Some beekeepers claim virtue in stacking the supers wet but I have found this to cause either fermentation, with a 'beery' smell in spring, or at other times, a residue of small crystals which caused more rapid 'setting' of honey in combs the following year.

I believe in making mistakes and learning from them. After trial and error I found that at temperatures of about 90 degrees F or above either the uncapping knife tended to 'drag' on the cells and deform the combs, or the combs collapsed when taken out. This is because beeswax softens and loses its tensile strength at such temperatures. It has to, or worker bees could not build their combs in the first place.

I also learned (the hard way), that there was a 'hot spot' immediately above the heat source, caused by upward convection currents as well as by radiation. A simple 'bridge' made of thin metal proved to be an efficient heat distributor, but even so I have found it wise to have wider spaces between the combs in the super immediately above the bulb, to allow hot air to rise. I also found a progressive gradation in comb temperatures over a range of about 15 degrees F. In practice this means that when the top combs are at about 70 - 75 degrees F the lowest combs are getting close to 90 degrees F, so that one may safely switch off and extract the pile progressively.

Friends have suggested that a purpose-made electric heater would be an improvement, but I have found ordinary light bulbs to be so effective and last so long, that I have never bothered. I did, early on, make the mistake of laying a bulb on its side on a wooden floor, but it burnt a hole, so now I rest the bulb in the central depression of a brick, or have it vertically supported in a holder.

Another use for this simple technique, is to keep OSR

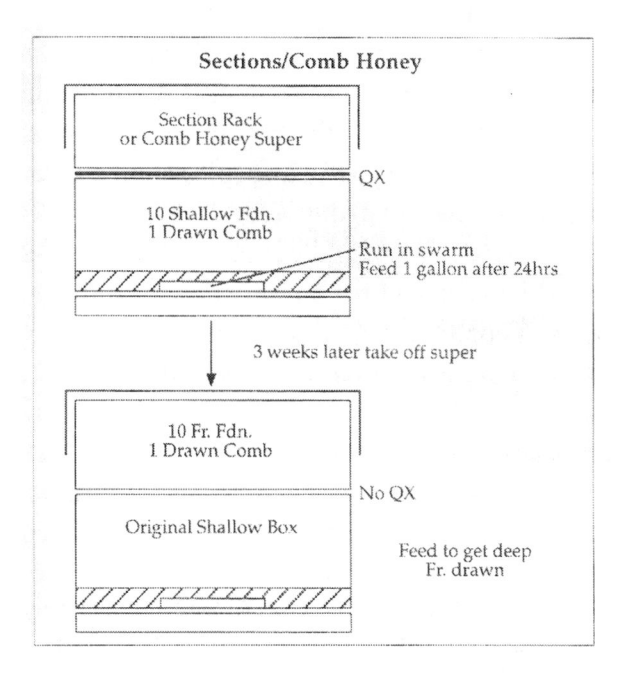

```
┌─────────────────────────────────┐
│      Sections/Comb Honey        │
│  ┌───────────────────────────┐  │
│  │    Section Rack           │  │
│  │  or Comb Honey Super      │  │
│  ╞═══════════════════════════╡ QX
│  │   10 Shallow Fdn.         │
│  │    1 Drawn Comb           │
│  │                           │── Run in swarm
│  │▓▓▓▓▓▓▓▓▓▓▓▓▓▓▓▓▓▓▓▓▓▓▓▓▓▓▓│   Feed 1 gallon after 24hrs
│  └───────────────────────────┘
│                │
│        3 weeks later take off super
│                ▼
│  ┌───────────────────────────┐
│  │     10 Fr. Fdn.           │
│  │     1 Drawn Comb          │
│  │                           │
│  ├───────────────────────────┤ No QX
│  │  Original Shallow Box     │
│  │                           │   Feed to get deep
│  │▓▓▓▓▓▓▓▓▓▓▓▓▓▓▓▓▓▓▓▓▓▓▓▓▓▓▓│   Fr. drawn
│  └───────────────────────────┘
```

supers warm to prevent granulation. I find that when brought home from hives in the field and stacked in heater piles of 6 - 8, they can be left for days and extracted at leisure.

Economical Use of Swarms

If the beekeeper reading this happens to live in an area not yet known to be infested with varroa, he or she may well regard swarms with suspicion, unless they arise from their own hives. If, as in Devon, you have lived with the pest for two or three years, you may once again welcome a swarm as an opportunity to carry out one or two interesting procedures. In any case a swarm is broodless for a few days, so any mites must be on the bees themselves and capable of being knocked down in 3 or 4 days by a couple of Bayvarol strips.

One effective technique for obtaining sections or cut-comb honey is shown in the diagram. The single drawn comb in the shallow box gives room for the queen to lay immediately and it is much less likely that the swarm will decamp. Swarming bees are gorged with honey and programmed to produce wax: the 10 frames of shallow foundation give them the opportunity. A large swarm will not fit into one shallow box, so it will spread through the QX into the upper box and cluster on the thin, unwired foundation provided. After allowing 24 hours for the bees to settle down and begin working, feed a gallon of syrup and leave alone for the next two or three weeks. The syrup will enable the wax foundation to be drawn into combs, and if there is a honey flow you should get 20 lbs or more of sealed comb honey or the equivalent in sections. In a really good flow you may well get a second box. Take these off before the comb surfaces become discoloured and add a brood box of combs or foundation (no QX). Feed syrup to help to establish the bees in their permanent brood box well before winter and replace the queen if you have reason to believe that she is old and likely to fail during the winter.

A good swarm can give you 10-20 sections or 20 lbs of comb honey, even in an average summer and if well fed afterwards, will still build up on deep combs well before winter.

Reinforcing

The diagram, illustrates an entirely different technique. The swarm is run into a shallow box fitted with drawn combs, but with a rigid QX below, so that willy-nilly, the bees have to go through it to cluster on the frames. A few will remain with the queen below the QX and all the bees feel that their queen is with them. Next morning, rotate slightly and lift off the shallow box, plus 98% or more of the bees, take up the QX, together with a small cupful of bees with the queen and replace the shallow box. At your leisure, invert the QX and gently push bees aside with a finger, so that you can inspect the queen. This 'mini-swarm' plus queen will nicely stock up an empty Apidea mini-nuc and you may keep the queen or replace her with a queen cell as you wish. The 98% of bees

Re-Inforcing

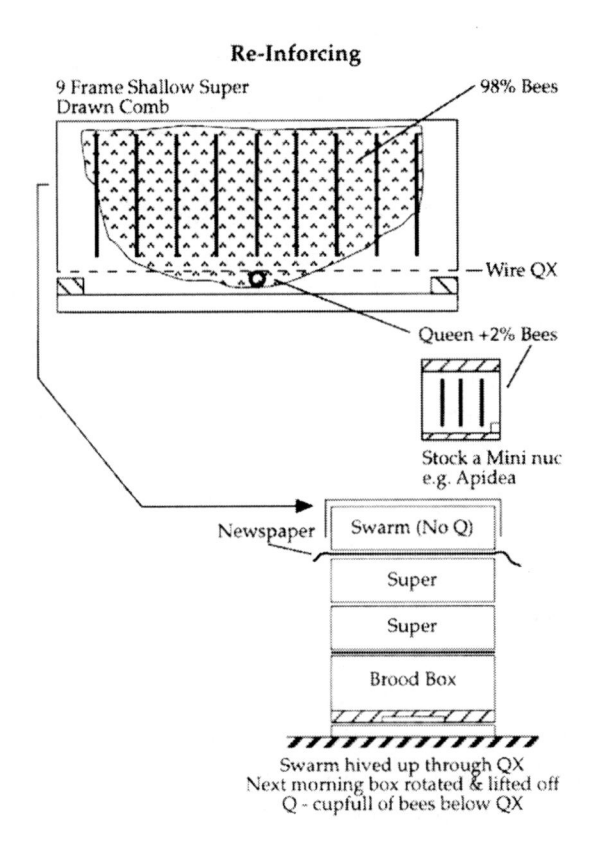

```
9 Frame Shallow Super          98% Bees
Drawn Comb
                               Wire QX
        Queen +2% Bees
        Stock a Mini nuc
        e.g. Apidea

Newspaper │ Swarm (No Q)
            Super
            Super
            Brood Box

Swarm hived up through QX
Next morning box rotated & lifted off
Q - cupfull of bees below QX
```

still in the box should be removed to another site or, loyal to their queen, they will hunt for her, find her and then form a swarm on the Apidea. This happened some years ago in my garden! In another apiary they can be used to reinforce an existing colony by uniting over a sheet of newspaper.

Induced Supersedure

This method of requeening without the difficulty of finding and removing the old one is based on New Zealand practices seen there twenty years ago. It was never claimed to be more than about 75% successful and trials over three years in this country (by D.A.R.G.) have only achieved just over 50% success. It is based on the fact that when queen cells are not accepted, the bees destroy them at a side; never at the end from which the queen will emerge. I always hesitate before using the word 'never' but in this case the usage is justified, as in over 100 cases of the introduction of a ripe queen cell, protected as shown in the diagram, the cell showed either natural emergence (98%) or no emergence at all (presumed due to internal damage). Not once has such a cell been damaged by the bees.

It seems that when a virgin emerges, the bees assume that she has come from a supersedure cell and may be accepted as such. If, however, the bees are very satisfied with the current queen, then they may not take much interest in the new one which, of course, also has to return safely from her mating flight.

An alternative found to be much more successful, is to introduce a protected queen cell between the frames of a super. Of course a top entrance must be contrived. Once the new queen is laying, she may be removed for sale or used elsewhere! The small amount of brood present will develop normally and the cells then used for honey storage; but, of course the old queen is still there below and attempts to have her superseded, by temporarily removing the QX and smoking new QX bees down into the brood box, have only been partially successful.

Wax Saver

Any variety of Miller feeder left permanently on a hive is a great economiser of time, in that when emergency feeding is needed, one has only to go round with containers of syrup. Such feeders can also be great wax savers. When at an out-apiary going through hives one comes across comb in the wrong place, often containing honey, one can scrape this off the hive tool into the feeder. By the next visit the bees will have licked it dry with no robbing and the wax can be

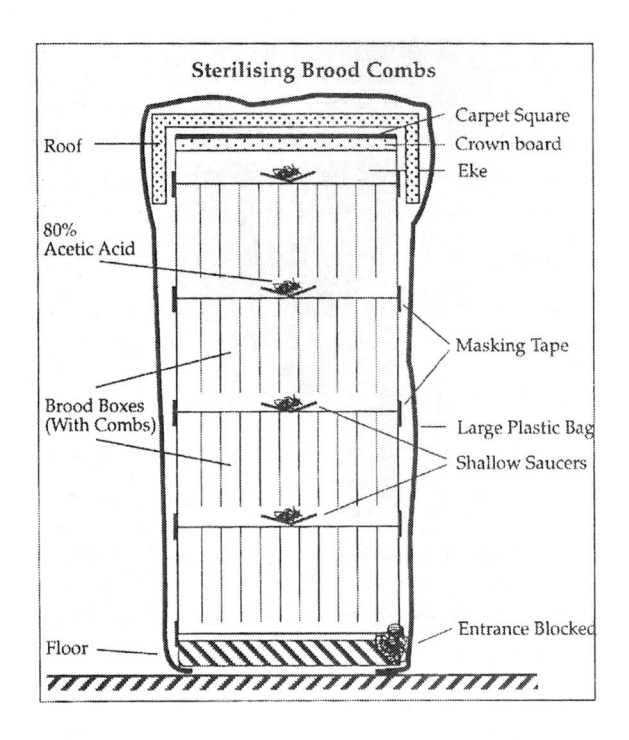

SECTION 2. ECONOMICAL BEEKEEPING 55

Sterilising Brood Combs

- Roof
- Carpet Square
- Crown board
- Eke
- 80% Acetic Acid
- Masking Tape
- Brood Boxes (With Combs)
- Large Plastic Bag
- Shallow Saucers
- Entrance Blocked
- Floor

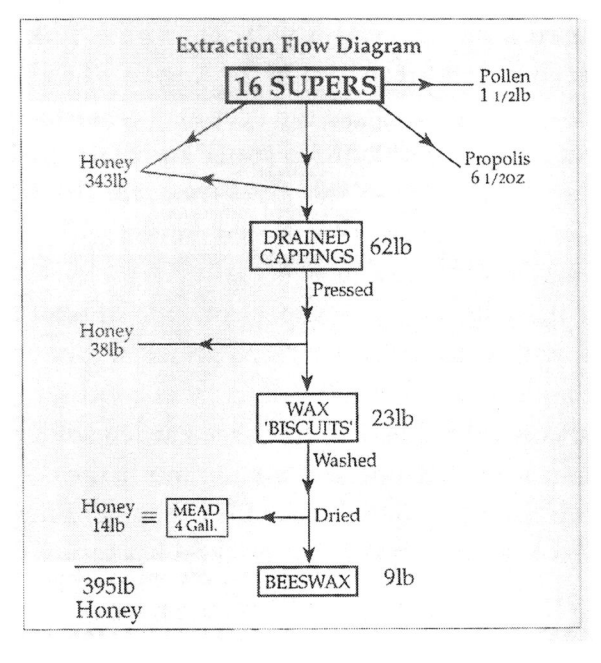

Extraction Flow Diagram

16 SUPERS → Pollen 1 1/2lb
→ Propolis 6 1/2oz
Honey 343lb
DRAINED CAPPINGS 62lb
Pressed
Honey 38lb
WAX 'BISCUITS' 23lb
Washed
Honey 14lb ≡ MEAD 4 Gall. ← Dried
BEESWAX 9lb

395lb Honey

collected. My own favourite is a home-made Buckfast-type, modified to use a hollow wood cylinder for bee access, with a short length of grey plastic drainpipe off-cut, topped with a glass or plastic disc, as shown in the diagram. The bees themselves found a new use for such a feeder two years ago. At my Powderham apiary I had been unaware of seven acres of OSR a mile away the other side of a wood. Having supplied insufficient super spaces the bees built a solid mass of honeycombs, completely filling the feeder!

BKQ 43
Autumn 1995

Maintenance of Equipment

Just as important as making new items is the maintenance of what you already have: to give just one example, the care of supers of drawn comb.

(*a*) In a bad year not all one's supers may be needed, yet a super unused for a whole year will deteriorate. I like to put such supers on hives for the month of August when bee populations. are still high, so that the bees may clean and service the combs. I also put an unused super above the QX when clearing supers to harvest. This gives room for the displaced bees and at the same time the unused combs are given their 'M.O.T.' and annual service.

(*b*) For the winter I stack supers outside in piles of 8 on iron rails, with gauze travel screens above and below. The first cold night normally kills off any wax moths, but as an additional control I welcome the large garden spider which usually finds its way in and sees off any moth; the occasional trapped bee also consumed I regard as the spider's perk-payment for biological control. This method has served me well for many years in the, mild climate of Torquay, when we have only the occasional frost in January and February. I would expect it to be even more effective inland and further north.

Perhaps I should not omit the more obvious routines, such as blow-lamping empty boxes and floors, sterilizing with acetic acid any brood combs not in hives, deep-freezing Apidea and other mini-nucs for 48 hours before plugging the entrance and storing them under cover. I also find it a great economy of time and trouble to have a few clean floors and queen excluders when going through the hives in spring. Instead of scraping a QX while on a hive, just replace it with a fresh one and do the cleaning up away from the hives.

Hive Products

The flow chart shows the production of honey, beeswax, pollen and propolis to be expected. from 16 very ordinary supers. The supers are warmed up in stacks of 6 to 8 and extracted when a domestic thermometer pushed into the combs of the top box indicates a temperature in the middle seventies. When uncapping over a coarse sieve on a large tub, the wet cappings drain while the extractor is spinning and the sieve may well hold the cappings from the 6 to 8 supers; if not, take a short break and do something else. If it is the end of the day, let the cappings drain all night before tipping them into a temporary storage bin and starting on the next pile of supers. Spun-out supers are put on hives in the garden to be licked dry before frames are checked for pollen (scraped out) and propolis (pushed off with a knife). The drained cappings are pressed or spun to get out as much honey as possible and then washed in cold water to obtain a "Must" of honey water of the right gravity for mead.

The washed cappings are then squeezed dry and spread out on a sheet in the greenhouse to get really dry, ready to be melted down to get wax blocks.

In the diagram I have calculated that the 4 gallons of rnead have used up 14 lb of the otherwise unobtainable honey left on the cappings. I count this as part of the crop, as I have a farmer who prefers mead to honey when I pay my rent in kind for hives on his land.

Harvesting Pollen

Pollen collected in a trap is comprised of pollen loads or pellets, of average weight 15 to 20 mg. (personal observations) which have to be removed and deep frozen every 4 or 5 days. If left any longer they may grow a surface mould. This is not surprising, as in the warm and humid atmosphere of a hive various organisms can rapidly establish themselves on a food so nutritious as natural pollen. What did surprise me at first was to find pollen vulnerability under English conditions exactly the same is in sub-tropical Queensland (Australia). I suppose that temperature and humidity conditions within a hive of bees does not vary by much anywhere in the world. Pollen stored by bees in cells for their own use later on has honey plus glandular secretions added to preserve it over winter, as well as being uncapped by wax. If for sale as pollen pellets (unrefrigerated) then careful drying is necessary, which then makes the pellets hard and difficult to reduce to a powder. Heat involved in the drying process may also damage some constituents of the pollen. Deep frozen, undried pellets stay soft even when frozen and, if required, can easily be transformed into a very fine powder of individual pollen grains by gentle pressure with the back of a large spoon or by a smooth surfaced wooden 'mushroom' as used by small-scale wine-makers to crush soft fruit.

When a lecturer at the National spoke of using a coffee grinder two or three times to crush pollen grains and expose their contents, he probably was referring to dried pollen loads. The actual pollen grains are as fine as flour already, as their average diameter is under 25 microns, i.e. less than one thousandth of an inch (much finer than ground coffee). In any case, pollen grains are extremely hard and virtually indestructible, having an outer case made of sporopollenin. Fortunately this outer case has pores and furrows through which enzymes in human (and bee) stomachs can leach out the nourishing contents. I write in some detail because of a widespread misunderstanding of these facts.

SECTION 2. ECONOMICAL BEEKEEPING

Pollen

Apart from trapping pollen, I also scrape it from super combs in autumn. As any beekeeper will know, there is a tendency for pollen to be stored in an arch in the central combs of the first super above the QX.

After extraction this is left behind and if ignored it will go mouldy, set hard and give much unnecessary labour to the bees next summer. I scrape it down to the midrib with a curved grapefruit knife or an old kitchen spoon worn sharp by years of use and store it in 2 litre ex ice cream boxes in the deep freeze until needed. Of course there is some wax with the pollen but I explain to customers that this is less than 2% of the weight of pollen, it is completely harmless when eaten and they are fortunate to have the bonus of 'pollen plus', as stored pollen has valuable bee secretions added to it which trapped pollen does not have. I make no claim, but dozens of friends have told me that just half a teaspoon of pollen-enriched honey (about 10% pollen) taken per day from February to July eliminates or at least reduces their hay fever. About 1 in 25 report symptoms of hay fever after the first dose, when I usually recommend no pollen for a week before returning to a much smaller dosage. What about the bees? Well, you have saved them a great deal of work dealing with mouldy pollen, but more importantly you have given them some wax-construction work to do at a time (next spring) when they are programmed to do just this by Nature. Having to fill drawn combs all the time is somewhat unnatural and it is always important to 'go along with the bees'. Giving them some wax-work to do in spring is just right.

Homogenising Pollen / Honey

Cooks will be familiar with the technique of mixing small quantities of powder with a liquid to produce a thick cream, before incorporation in large amounts of liquid. I do this with a small amount of warm liquid honey added to the pollen before attempting to homogenise the main honey / pollen mixture. This cream is then added to warm bulk honey in a stainless steel bottling tank and thoroughly mixed with a plunger before filling honey jars. Failure to reduce pollen pellets to a fine powder results in the pellets floating to the surface in each jar.

If using pollen scraped from super combs after extracting, the task is more laborious, as the stiff cell content plus some wax have to be thoroughly mashed with warm honey. I then spoon it on to the built-in steel filter of a 56 lb honey tank and scrape it through with a spoon. Much of the pollen/cell wall mixture remains above. I let a few discerning customers have this, or eat it myself.

Harvesting Propolis

I regularly scrape propolis from super frames after extraction, as well as from rebates in boxes; I never collect it from floors. Apart from this, I use a plastic grill under the crownboard after mid-June. In late August I deep-freeze this for 12 hours to embrittle the propolis and flex to disengage it. I find the best propolis has a deep chestnut red colour. From time to time some bees build a barrier of propolis in a wide entrance and of course, the name itself stems from this habit. My bees in Central Africa did this far more than my bees in Devon; possibly to exclude prevalent black/red coloured beetles often trying to enter hives, a nuisance we seem to be free from here.

I market propolis in plastic bags direct to customers at £2 per 20 gm but have also sold much larger quantities to violin makers (in Norwich for example) who make a varnish from it. For personal use I steep about an ounce (30 g) in a small bottle of gin (150ml) for a couple of weeks, shaking occasionally, and decant the coloured liquor for use against sore throats. For young members of the family, a couple of drops on a sugar lump seem to give relief. Older family members just gargle with small amounts and spit out or swallow as they wish.

Mead from Cappings

Mention has already been made of draining pressing or spinning cappings to get the maximum amount of honey from them. After all this, the apparently 'dry' mass of cappings does in fact still contain far more honey than most people realise. My practice is to add about 20 lb/ 10 kg of such cappings to 3 1/2 gallons / 16 litres of cold water in a large plastic drum (normally used for the aerobic stage of wine or beer fermentation). With sleeves rolled up, the cappings are hand-crumbled and the gravity adjusted by adding water or more cappings until a hydrometer shows a suitable figure (1.090 to 1.105 to personal taste).

By tilting and decanting about 95% of the liquor can be poured off. The remainder is hand-squeezed to cricket ball sized lumps to get out most of the rest. Though still moist relatively little honey remains - the final liquid residue is mostly water. The wax balls are dried and finally melted, while the liquor, or 'must', is fermented for 3 - 5 days in the same large drum with a wine yeast. When the original vigorous aerobic fermentation has abated, the 'must' is poured into a large 4 gallon glass vessel (or several smaller ones) fitted with 'gloppers' and allowed to ferment for 6 weeks, by which time bubbling will have ceased and the liquor can be racked off and cellared in bulk for six months before bottling.

Bottles of mead find ready acceptance as part of apiary rent. On the basis of one bottle of mead needing only 1/2 lb honey, one gains several extra pounds of honey, plus additional goodwill.

Quite apart from these points, there is the fun of running a bee-produce stall (plus observation hive at a church fete, for example. Sometimes in August this coincides with a bumper crop of ripe plums from my garden and I have a ready sale for 1/2 lb bags of my special 'bee-pollinated plums'. I also give a button to any girl visiting my stall; the smile on the face of a 9 year-old wearing a tooth brace on receiving such homage just makes my day!

Mead from Cappings

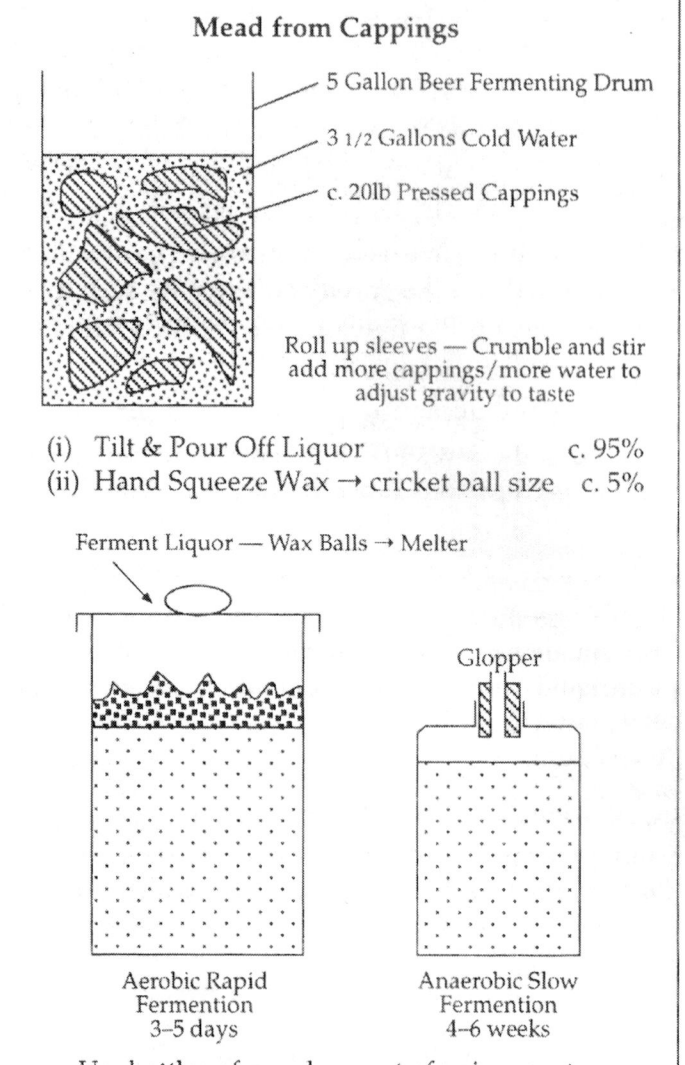

- 5 Gallon Beer Fermenting Drum
- 3 1/2 Gallons Cold Water
- c. 20lb Pressed Cappings

Roll up sleeves — Crumble and stir add more cappings/more water to adjust gravity to taste

(i) Tilt & Pour Off Liquor   c. 95%
(ii) Hand Squeeze Wax → cricket ball size   c. 5%

Ferment Liquor — Wax Balls → Melter

Glopper

Aerobic Rapid Fermentation
3–5 days

Anaerobic Slow Fermentation
4–6 weeks

Use bottles of mead as part of apiary rent
eg - for a hive apiary instead of 15lb honey many farmers prefer $1/2$ doz bottles mead + 8lb honey
(1 bottle mead = $1/2$lb honey, you gain 4lb honey + extra goodwill)

SECTION 2. ECONOMICAL BEEKEEPING

Package Bees

Some uses of swarms were quoted in the last issue. Here is another.

Throw the swarm on a white sheet or flat piece of plywood in front of the entrance to a brood box with a rigid queen excluder placed between floorboard and box; also of course a crown board. The brood box should preferably contain drawn combs on which the bees may cluster. If short of drawn combs I have found that frames of foundation serve but it does help to have two or three drawn combs in the centre. The actual cluster will reach down below the QX, where the queen will be.

Next morning the brood box may be rotated gently and lifted, leaving a small clump of bees below the QX (for stocking a mininuc). The queen-less cluster may be given a queen cell, or a young queen, or used as package bees. It is best not to feed until after 24 hours, to lessen the risk of the bees decamping.

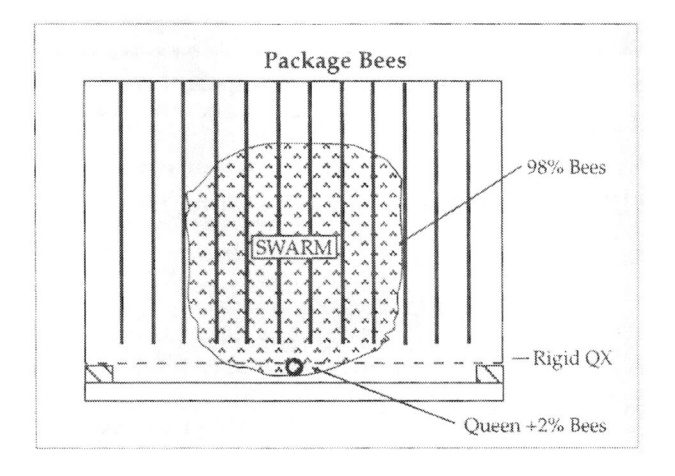

Package Bees

98% Bees

Rigid QX

Queen +2% Bees

SECTION 3.
A TOP-BAR HIVE

BKQ 44
Winter 1996/7

Here is something completely different; a hive using no frames, no supers, no queen excluder, no sheets of wax foundations, envi-ronmentally friendly and costing very little, attractive in any garden. Based on the principle used by the Greeks 2,000 years ago in their basket hives with sloping sides, but forgotten for centuries. They realised that when side walls are inclined to the vertical by 14 degrees or more, the bees do not join their combs to them as they do when the sides are vertical. Economical in use, involving no lifting of heavy boxes, yet with all combs readily available for inspection at all times, whether for checking queen cells, re-queening or harvesting. A similar hive about one third the size of that shown can be used as a nucleus; in fact, a miniature version in polystyrene with three or four tiny combs is in use by the Germans as a queen mating mini-nuc.

Natural Material

As can be seen from the photographs, the basic materials are the bark-lined strips which sawmills take off a tree trunk to square it up before sawing it into planks. Of little commercial value except to gardeners, who sometimes use this natural produce to make plant containers. The main body of the hive is made from three similar lengths, each 3ft long by 8 to 9" wide, with two end pieces cut into the shape of a trapezium measuring 12" at the top and 8" at the base. Even these measurements are not invariable, as no frames of constant size are involved and the top bars can easily be cut to fit whatever size you choose, as the bees make their own bee space, so long as the sides are inclined at an angle of at least 14 degrees to the vertical. If the sides are vertical, as in an orthodox hive, the bees will attach their combs to them and so these can no longer be lifted out at will for inspection.

The top bars are cut from thin strips of wood 1 3/8" (3.5 cm) wide and fit into a rebate cut into the side pieces. The model illustrated (in use at Cockington Apiary, Torquay) has 25 top bars, each 12" (30.5 cm) long. In order to get the combs built correctly one can provide starter strips of wax 1/2" deep. In Central Africa over 40 years ago, I had no access to wax foundation so used a carpenter's scribe to cut a groove down the centre of each bar and ran melted wax down this groove. I thought I had invented this idea, but years later read of Thomas White Woodbury, of Exeter, doing much the same in 1859!

Top bar hive showing entrance at one end of the hive

Top bar hive with lid removed

Removing combs - this is much easier once bees have been raised in the cells

Beekeeping is like this and one often 're-invents the wheel'. In practice the bees did sometimes build their combs at an angle across the bars and complicate my work, so today I prefer to stick thin strips of foundation into the grooves. Once you have such a hive, you can start up a second one by using combs of bees from the first, but to get going in May or June, it is easy to run in a swarm. As with any hive having combs the 'warm way' the bees will store honey at the end furthest from the entrance. This is where comb honey (with no brood) can be harvested at any time by cutting out and then replacing the wooden strips, taking care to leave 1/4" of wax in place for next time. Comb renewal is simple, as the oldest combs are always nearest to the entrance and after removal the others can be pushed in to close the gap.

Comb Inspection

It might be expected that bees would propolise the bars heavily and make it difficult to move them. In practice this does not seem to happen, so long as the bars are pressing close against each other to make a continuous area of wood, like a cover board. A smear of vaseline along the rebate helps to make it easier to push bars along to close up any gaps. With no supporting wire or frames newly built combs are fragile and must be held vertical at all times, never at an angle. Once a comb has been bred in it is much stronger and can be handled much more readily. One soon learns the knack of doing this, as shown in the photograph.

Honey Harvest

Obviously the natural combs can be cut out and used (or sold) as they are, or chopped up to fit into standard 8 oz plastic containers. It would be very difficult to centrifuge these combs as we normally do. In tropical Africa 40 years ago, I used home-made hives, but with vertical sides, these were so much harder to work. At no time did I have an extractor, yet regularly bottled and sold large quantities of liquid honey at 3/6d a pound, having no competition except from imported honey. This price (17p today) may sound very modest, but honey in the UK was only 2/6d a pound at that time. Of course I also sold honey combs by weight, preferred by many customers, including one or two living in Devon today but they have to pay much more now!

Feeding

In Africa I never needed to feed bees and even in the rainy season they were able to maintain themselves (but of course, gave no surplus). Here in Britain one can easily feed at the rear of the hive, using bakers' fondant. An ingenious beekeeper can make or modify a feeder for use with syrup if desired.

Roof

The hive in use at Cockington (made by Roger Kirk to my specifications) has a roof made of thick plywood framed by wooden bars, having a waterproof felt top. Because of its large area it is securely tied on for winter .

SECTION 4.
SWARMS ARE PRECIOUS: WHY NOT TRY A BAIT HIVE AND SEE WHAT HAPPENS?

BKQ 57
Spring 1999

Every year there are many reports of swarms hiving themselves in empty brood boxes, or even in stacks of honey supers, and there is no doubt that boxes that other bees have lived in are most attractive to bees looking for a home. If you have an empty hive, with frames of used combs, in the garden, you may notice bees going in and out as if curious. Should the number of bees doing this increase considerably over two or three days, they are probably scout bees sent out to look for a home by a colony about to swarm. The scout bees may even come from a swarm of bees hanging unobserved in a tree up to half a mile away. Within 48 hours or less the swarm may arrive.

Why not Prepare a Bait Hive and see What Happens?

You will need a brood box (preferably British National size) in any case, so get it ready as soon as possible and fit it out with ten frames of foundation and one frame of old but clean, dark brood comb (to attract the scout bees). Even better would be a secondhand four or five frame nucleus box with frames of combs that have been bred in, and a small entrance smeared around with wax and warmed propolis scraped from one of the old frames. Place well above the ground, about 5' - 6' (2m) up; this is then very likely to attract a swarm during June and July.

When occupied, the bees can be shaken into your 'proper hive' and the trap set up again. Transfer just one of the old combs plus bees and replace it in the nucleus with a frame of foundation.

What are the snags? If you leave any space in the the bait hive not occupied by frames, the bees will choose to fill it with new, wild combs, perhaps built at an awkward angle, rather than use the frames provided. There is also the risk that wax moths will get to work on any old frames of comb and make a horrible mess of them, but these risks have to be taken, for a few weeks only.

What are the ethics of this? In law bees are wild creatures temporarily housed, but when they escape from an apiary they are considered to have regained their freedom, unlike cattle or sheep, hens or geese which are domesticated. Among beekeepers themselves, ownership of a swarm is usually conceded if a beekeeping neighbour says that he saw a swarm from one of his hives fly towards your premises, or if the queen is marked in a distinctive way. In practice this is rare, and a swarm may have come from a church steeple or a hollow tree.

(I see nothing at all wrong with putting up bait hives. It is a common practice in many countries (though illegal is some like Switzerland) and I remember a pleasant afternoon I spent with Job Pichon in Brittany doing the round of his baits. It took me some time to see them for they were often quite high up or hidden in the trees. Swarms are extremely valuable today - better that they end up in a bait hive of a beekeeper who is going to look after them and treat them rather than become a nuisance to the public should they enter a wall space and thus end up being destroyed. Editor.)

Using a Swarm to Plump up a Colony

Plumping is stocking up a skep or small wooden box (about 4 gallons or 20 litres) and placing it alongside a honey-producing stock and feeding it to get natural combs built. After 5 or 6 weeks, working from the rear, just pick up the skep or box and place it on the other side, when bees flying from it will return to the original site and be accepted into the main hive alongside.

A further refinement is to turn the moved skep so that its entrance is in a different direction, and work it back to the position shown over the next week. When the skep bees are once again flying strongly, the manoevre may be repeated. A reinforcement of several thousand young bees in a honey flow should help to give you a record crop, and the skep can be fed in autumn and brought through the winter with an

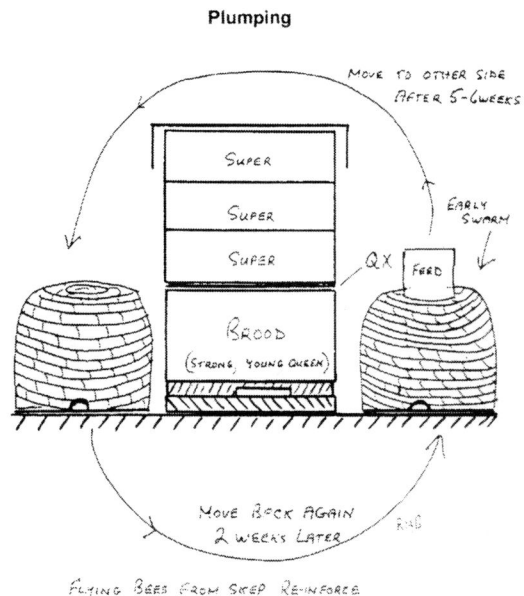

Plumping

Flying bees from skep re-inforce honey producing stock.

improvised metal roof. As a boy in the early 1920s I remember Mrs Crick, a skep beekeeper in our village, having 5 skeps over winter in her garden, with an assortment of old metal pans to keep them dry: not exactly beautiful, but they kept out the rain.

These two items on swarming are from HONEY-BEE SWARMS, by Ron Brown, just published by Venture Press for £3. The 30 page booklet is a mine of practical information on how to utilise swarms for optimum profit in the apiary as well as containing humorous, entertaining stories of the author's swarm exploits. A boon to anyone who wants to make the most of their bees during these difficult times. Editor.

SECTION 5.
VARROA: A REVIEW

BKQ 30
Summer 1992

As is well-known, Varroa jacobsoni Oudemans is an external parasitic mite of pinhead size, adapted over a very long period in the Far East to live on Cerana bees, slightly smaller cousins of our Mellifera bees. A wise parasite does not kill its host, and the Cerana/Varroa equilibrium is based on two main factors:

(a) the shorter incubation period of Cerana pupae, so that successful Varroa breeding is virtually confined to drone brood, and

(b) the 'grooming habit' of Cerana workers, which remove mites from themselves and each other, like monkeys do fleas.

About 70 years ago, Varroa transferred successfully to colonies of Mellifera taken to the Far East to 'improve' beekeeping there. From the point of view of the mite, these bees provided a 'soft target' and what was just a minor nuisance to Cerana has proved a killer to Mellifera.

Arrival in Europe

The pest arrived in Europe about 20 years ago in at least two ways:

(a) From the Russian Far East via hives moved by rail to the Moscow area, then to the Russian zone of occupied Berlin by soldier beekeepers allowed to bring their bee nucs with them for recreation.

(b) By the importation for research purposes of Cerana colonies into Central Europe (no names!). Germany has had Varroa now for over 15 years and France for 10.

We know that it has existed on the Channel coast for at least five years and, in my opinion, it probably arrived in Devon by a swarm or cast on a 'roll-on', 'roll-off' large vehicle via Plymouth. The swarm probably escaped when the vehicle parked in the Okehampton/Bow/ Crediton area, recently found to be the larger of two 'epicentres', the other being in the Southampton area. Since swarms usually have relatively light infestations, the pest has been slowly and quietly multiplying and spreading via hives taken to Dartmoor/Exmoor for the heather.

Thanks to the vigilance of Torbay beekeepers, a very low infestation was discovered at Cockington on 4 April 1992, shortly after the 'all clear' letter had been received from Luddington in respect of floor debris conscientiously sent in, in response to well-meaning exhortations from various authorities. As is well-known a real search then began and much heavier infestations (some already lethal) were discovered in Devon and Hampshire, and elsewhere in Southern England.

Cross-country Spread

The pattern on the continent (and in the USA) has been of a slow, natural progression of varroa via the bees themselves combined with a rapid spread by beekeepers moving hives, for pollination and other reasons. Many areas escaped for several years, but in the end the problem became universal. In Britain it is likely that this pattern will be repeated; for a number of reasons, rather more slowly, and Scotland could well be free until next century (save for movement of hives by beekeepers).

The shorter breeding season of bee colonies in cooler northern countries tends to slow down the development. For example, the mite reached the south eastern area of Finland over nine years ago, but has made scarcely any progress further north. However, the whole of Scotland (except the Shetland Isles) lies further south than any part of Finland, so Scottish readers must not be too optimistic.

Experience Elsewhere

No country once infested has yet been able to eliminate the pest, and attempts to do so by the large-scale destruction of infested stocks have always failed, because other colonies thought to be free were not. On the other hand, methods of control available today did not exist several years ago when varroa was spreading across Germany and later, France. We do not have to repeat relatively ineffective control methods but can proceed directly to the modern methods being used successfully across the Channel.

Germany: Perizin/coumophos has been used for over four years now. A 10ml bottle is diluted with half a litre of water to give a 2 per cent solution, sufficient to treat 10 strong or 20 weak colonies in Spring and early autumn (at times when there is no honey flow). The liquid is dribbled over seams of bees between brood combs, without taking out the frames. Bees drink it and the action is systemic, being lethal to mites that attempt to suck heamolymph. A government subsidy reduces the cost to beekeepers from Dm.lO to Dm.1.67 per 10 ml Perizin.

France: Here Apistan/Fluvalinate has been used for almost as long. It is applied via lanieres or strips of material soaked in fluvalinate and suspended between combs 3 & 4, and 7 & 8 of the brood box for six weeks in February/March and again mid-August to end of September. Apistan is a contact acaricide, not affecting mites in capped brood cells (as 80 per cent of them are when brood is present) so that enough time has to be allowed for mature mites to emerge.

Austria: Treatment here is very similar to that in France, except that Fluvalinate and Bayvarrol are used in alternate years.

Belgium: Here a few weeks ago we attended a large meeting of Flemish beekeepers at Zutendaal and saw the thin strips of wood soaked in fluvalinate which they are preparing and using themselves, with great success. One new point stressed by them was the need to treat before the end of July (immediately after removing the honey supers). They have brought forward the date from early autumn after having experienced the sudden collapse of apparently strong stocks, as young mites emerged in huge numbers from summer brood and attacked the bees, having little or no fresh brood to enter.

Perhaps the success of these chemical control methods is best illustrated by what beekeepers are grumbling about. Is it Varroa? No, it is the low price their honey makes and the difficulty of selling sunflower honey (France) and oil seed rape honey (Germany)!

Four Years to Kill

Year 1: From an initial attack by 4-6 mites, which may have reproduced in just six drone cells, there could be 30-50 mites present at the end of the first summer, with no visible sign of trouble at all. These will overwinter on worker bees, feeding very little, and by early spring with mortality of around 10 per cent there will be perhaps only 30-40 present.

Year 2: Rather surprisingly the overwintered mites (all mated females) often leave the first worker brood of spring almost untouched and prefer to wait for the first drone brood (late March/ April) before entering the cells just before they are sealed, feeding heavily on the larvae for 2-3 days and then laying up to six very large eggs at thirty hour intervals on the developing drone pupae. The eggs produce adults in 7-8 days and the growing stages of the mites suck the life blood of the bee pupae, before maturing and mating within the cell and then emerging with the adult (deformed) bee. Unmated females die in the cell. One of the first two eggs laid produces a male, which mates with its sisters and then dies in the cell. This flagrant abuse of the male sex is one of the nasty habits of the spider or Arachnid family, to which the eight-legged pest belongs. At the end of the second year there may be perhaps three hundred mated females overwintering on live bees (and transferring to others if or when their particular host dies). So far, very little worker brood has been affected and most beekeepers would have noticed nothing, but unknown to them, the other hives in their apiary will now have 4-6 mites, transferred by drifting bees or visiting drones.

Year 3: By early summer there may be insufficient space in drone cells for three hundred females and some mites will be breeding in worker cells. With two days less for their life cycle, only two mature and mated females per cell will then be produced (instead of four in drone cells), but on the other hand about 25 per cent of overwintered mothers will repeat their performance, after a few days' rest on nurse bees. Some 5 per cent will do even better and produce a third brood in the same year. By now, any observant beekeeper will notice occasional deformed bees on the alighting board. The leave-alone 'bee owner' will carry on putting supers on and taking them off, and in a good year, may still harvest a small crop and be completely unaware of the time-bomb due to explode next year.

Year 4: With overwintered mites now numbered in thousands rather than hundreds, population pressure will drive them to breed mainly in worker brood, and even the 'bee owner' (should he ever look inside a brood chamber) will notice the massive 'pepper-pot' effect, with dead larvae and neglected brood. At this stage it can resemble foul brood, and our friend may even 'phone the local Bees officer.
Alas, the condition is now fatal, and at the end of summer,

even the strongest colony, so affected, will break down. The end is accelerated by the additional hybrid vigour brought about by cross-mating in cells with more than one female mite, opposed to the brother/ sister in-breeding of earlier generations. By now, all other stocks in the apiary will be infested, and visiting drones will probably have transferred the mites to hives of unsuspecting colleagues who keep bees up to five miles away.

The mite can only survive for 3-5 days outside living colonies, but robbing bees in the last week or two can carry the pest back to their hives. However, by September, the only task left is to shovel dead bees and dead mites on the compost heap or bonfire, fumigate the combs and be ready to start again next year, a sadder, wiser beekeeper.

Control by Management

On the continent several years ago the emphasis was on making up varroa free nuclei with young queens and building them up rapidly to give a honey crop before varroa infestation levels became too high.

This was supplemented by insertion and removal of drone combs, trapping perhaps up to 50 per cent of the mites in the process. My impression in more recent visits (over the last two years), is that such labour-intensive methods have now been almost entirely replaced by chemical treatment with Perizin or Fluvalinate.

Folbex V. A. has also fallen out of favour.

Another method used by some commercial beekeepers, mostly in Japan and Russia but also, to a limited extent, in Germany, is heat treatment, based on the fact that varroa mites are more sensitive to heat than bees. Very careful thermostatic control by hot air cabinets is necessary, as bees are badly stressed at temperatures only marginally higher than those used to dislodge the mites. Obviously mites in sealed brood cells are unaffected.

Contrariwise, a colder climate shortens the brood season

(as in Finland) and this can be partly achieved by using gauze panels in floors and no crown boarding. The bees seem to catch up rapidly once breeding does get under way, and I have heard German beekeepers remark that the workers for rape in April are already in the hive from the previous autumn.

Chemical Residues

It is very clear that the problem here is public perception of a possible danger rather than the reality of the situation. Our modern technological ability to detect infinitesimal traces of almost any substance has fuelled public concern over amounts far below any possible danger levels. R. Borneck, President of the World Beekeeping Association and a scientist in his own right, has researched thoroughly into the residues left by Apistan, Perizin and Folbex V. A.. Levels found in honey ranged from 0.003 part per million for Fluvalinate to 0.26 p.p.m for Perizin. Levels in beeswax were slightly higher, yet in wax melted from worker brood combs in which fluvalinate strips had been continually present for three years and renewed annually, the maximum residue was 4.9 p.p.m. After rendering, pressing and converted to wax foundation for sale, even this sample had only a residue of 0.8 p.p.m. British environmental health officers have confirmed that values such as these are a very long way below any possible danger level.

However, the public at large reacts adversely to the word 'chemi-cal' ignoring the fact that even pure water is a chemical made up of hydrogen and oxygen! For the sake of good public relations, if for no other reason, we should insist that no chemical treatment be given during a honey flow when bees are storing in supers.

Long-term Research

(a) At Oberursel Research Station in Germany (Prof. Koeniger) last summer I saw a large-scale experiment involving hundreds of mini-nucs, in an attempt to obtain a hybrid bee with a shorter breed cycle, resembling that of the Far Eastern Cerana bee in which varroa can only brood in drone cells. They were crossing Capensis bees from Cape Province, South Africa, with local Carniolan bees and obtaining some measure of success.

(b) At Bures-sur-Yvette in France some basic research is in progress to identify and synthesise any pheromone by which varroa mites are more strongly attracted to drone brood. Obviously any such pheromone could be used in some sort of bee-proof varroa trap.

(c) Encouraging 'natural' resistance.

Since I first saw varroa in Brazil (1978) with Dr. Koeniger and Prof. Morse, I have followed its progress in the Americas with great interest. In Brazil the infestation of so-called 'Africanised honey bees' has remained at low levels, and beekeepers generally do not treat against the mite. In Uraguay the varroa infestation was assessed at only 5.5 per cent of colonies, very similar to the level in South Vietnam (Mellifera in both cases). The mechanism is partly attributed to the shorter time period of sealed brood, but perhaps even more importantly, to the 'grooming habit', i.e. the way in which Cerana, Capensis and Scutellata honey bees literally pull the mites off each other. Dramatic pictures recently published in an Italian bee journal show mites with broken legs and cracked body shields after rough treatment by the bees. (J. Woyke).

(d) A further resistance reaction (noted by Prof. Drescher and Prof. Koeniger), is the habit of either opening up and clearing out infested brood cells, or even thickening the cell cappings to prevent emergence of crippled bees plus mites.

(e) It has been observed that an occasional colony of Carnica honey bees has shown a natural resistance; this was discovered by the presence of mutilated mites on the hive floor. One worker assembled 13 colonies showing this tendency (out of 700). Live mites were introduced to one of these and in a short time dead mites (some mutilated) were dropping as debris. From the autumn of 1990, of these 13 colonies overwintered without protective treatment, just one remained at a static level of infestation but the others needed treatment. There is the possibility that breeding intensively from such a stock could hasten the evolution of strains of bee with a natural immunity.

The Way Ahead

Diagnosis. In the next three to five years, accurate diagnosis is going to be extremely important, as there is no point in the chemical treatment of varroa-free colonies, yet early warning of even a low infestation is vital. As was demonstrated at Cockington on 4th April (and again for the BBC on 22nd May when just one mite was discovered after uncapping over 200 drone cells) the most sensitive test is by uncapping drone brood early in the season, using a 21-tined honey uncapping fork, lifting up 25-30 drone pupae at a time.

This test is most effective early in April, typically when opening up the brood chamber thoroughly for the first time. It helps enormously if a comb having its lower one third composed of drone cells (or drone foundation) be introduced at the edge of the brood nest in mid-March, when carrying out the first partial check to establish whether queen-right and what food is there. I have done this for the last ten years, partly to preserve the integrity of good worker combs. Bear in mind that with a very light infestation in March or early April, almost one hundred per cent of the mites will be on the drone brood, and a smoke test is likely to be negative.

Treatment. As a swarm consists of broodless bees, also likely to be much more lightly infested than the stock it comes from, it should be treated with one fluvalinate strip for two weeks immediately after hiving. Any mites present must be on the bees for the first few days, and so very vulnerable.

On the continent, all beekeepers agree that chemical treatment of established colonies should be carried out at the same time, not only all the hives in the apiary but also in the same area, otherwise rapid re-infestation will render any treatment ineffective. To this end it is vital that all beekeepers in a region get to know each other and cooperate, even if previously some were members of a bee association and some were not.

Controlled research in France on reinfestation at a distance of 500 m. has shown that drones visiting colonies with queen cells or virgins are the main natural vectors. The robbing out of dying colonies would be another prime cause, and over much shorter distances the drifting of flying bees from one hive to another.

In France the official recommendation is to suspend fluvalinate inserts between combs 3 & 4, and 7 & 8 in the middle of August and leave them in for six weeks. This treatment to be repeated in February/March, i.e. before the honey flow.

I strongly commend this as the method to follow.

The Next Cloud on the Horizon?

Any time now some writer on bee problems is due to make our flesh creep by speaking about the tropical bee pest *Tropilaelaps clareae, a dreadful cousin of Varroa jacobsoni.* Worry not; in a recent communication from Norman Rice (Australia) describing a visit to help Pakistani beekeepers, sponsored by the Pakistan government, he says:

> *"I saw these tropical mites crawling in and out of cells and running around on combs. They feed on the haemolymph of developing bee larvae/ pupae. I saw many worker pupae with holes chewed into their heads, dead or dying and having to be removed from cells by workers. If a colony is broodless for a couple of days, the mites die as they cannot feed off mature bees. A local management system is to move bees to a situation where they will become broodless. The problem is then cleared up until colonies are re-infested via other bees working in the area."*

At least these mites should not trouble beekeepers in

Europe.

Sources

"Apitalia" Jan. 1992 *"L'Ape Nostra Amica"* Feb. 1992 *"La Sante de I' Abeille"* Various issues.

Personal communications while visiting France, Belgium, Germany, Austria and other countries.

VARROA:
YOUR QUESTIONS ANSWERED

BKQ 31
Autumn 1992

Following Ron Brown's article in the Spring issue of The Quarterly, there have been several requests for more information on the subject - especially since the licensing of Bayvarol by Bayer, and the publication of MAFF's new publication on varroa.

1.
When can Bayvarol best be applied to colonies so that it has the maximum effect in controlling the mites - and with less danger to the honey?

Without any doubt, as soon as the honey has been taken off, at the end of the main flow. Not later than mid-August for at least two reasons, viz:

(i) When the brood nest is rapidly reduced in late summer, large numbers of mites emerge and cannot find larvae on which to breed, so of necessity attach themselves to and feed on adult bees.

(ii) Possibly due to 'hybrid vigour' arising from cross-mating in brood cells with more than one mother mite, the late July generation of mites seems to be more dangerous. We

heard in Belgium this summer of the sudden breakdown of apparently strong colonies, before they had even put in the fluvalinate strips.

2. Will one treatment per year with Bayvarol be enough for complete mite control?

For control, yes. For complete control, use strips also in February/ March, as many do in France. Possibly this is not so necessary here yet, as varroa still exists in a series of isolated pockets and re-infestation problems are less than they will be in several years time.

3. Why should the strips be left in the hive for six weeks? The longest brood cycle within the colony is only twenty-four days.

The longest relevant part of the brood cycle is in fact only 14/15 days, but the safety margin given by up to six weeks is advisable for several reasons, viz.
   (i) Bayvarol is a contact acaricide, and as such is spread more slowly through the hive than Perizin, which is a systemic and circulates more rapidly by food-sharing. Perizin, used in Germany, is diluted x 50 and 40ml poured on two seams of bees between brood combs. The bees drink it, and pass it to others.
   (ii) When applied in mid-August, the six week period covers most possible re-infestation up to the end of September, by which time there is little worker drifting and almost no visiting drones.
   (iii) Bayer have kindly supplied me with test results giving the average daily knock-down per hive of 16 hives in two apiaries (Germany), also of 15 hives in Holland. In the former the figure dropped from 225/day 1, to 5/day 18: in the latter from 400/day 1, to 5/day 21. In neither case did the figure quite reach zero, probably because of minor re-infestation.

SECTION 5. VARROA

Nevertheless, it was estimated that total knockdown was between 99.7% and 99.8%.

Results very similar to these have been obtained in France with Apistan, and the French authorities also advise 6 weeks.

4. Are beekeeping organisations - at least locally, trying to synchronise the treatment of varroa-infested colonies within their areas? Such schemes would surely be more effective.

Yes, we in Devon certainly are, both centrally and locally. For example, we held a very well-attended public meeting at Bicton College especially for non-members, to make this point. The registration of all beekeepers would help, as on the continent.

Even democratic little New Zealand has compulsory registration.

5. Considering the high cost of Bayvarol, is it possible:
(a) like on the continent with Apistan strips, to remove the strips from the hive after treatment and seal in a tin for the next year?
(b) when using the strips for diagnostic purposes, after leaving them overnight in one colony move them onto other colonies on following evenings until all stocks in an apiary/area have been checked for varroa?

(a) Based on the low vapour pressure of flumethrin, storage in a sealed package in a deep freeze should be very effective. It would be fairly easy to establish this by comparing diagnostic knock-down effects of strips on similar hives in an infected apiary.

(b) Yes, we have been doing just this at Cockington, and elsewhere in Devon, these last few weeks, with successful results.

As a corollary we have made comparative trials with tobacco smoke. The same hives which produced 0,1 or 2 mites after tobacco smoke gave 23 - 30 after one Bayvarol strip overnight. In practice it is probably enough to test one hive per apiary. If positive then treat all hives there. If negative, test one more (out of 10 to 15 perhaps) and if still negative press on to the next apiary.

6. Cheaper methods of varroa control e.g. formic acid and lactic acid are used on the continent and such substances are easily brought into this country. How can these chemicals be used and what safety precautions are necessary?

Formic acid can be dangerous to bees and humans. Think of it as being like acetic acid but more vigorous in all its reactions. It has the advantage that its vapour can penetrate into brood cells, but most reports quote its efficiency as being well below Bayvarol. Normally, 40ml of an 80% solution is poured on to an absorbent mat then pushed into the entrance, on the floor. Unlike Bayvarol its action depends on evaporation, which is too slow below 10 degrees Celsius and too fast above 25 degrees Celsius.

Lactic acid applied by spray to the face of combs is still being researched, but reports seen so far indicate that it also is less effective than Bayvarol. Both are superficially attractive in that they are 'natural' products.

So far as cost is concerned, it would be an absurdly small item in the national budget to subsidize Bayvarol, as the German Government does Perizin. This could be amply justified on environmental grounds, enhanced seeding of natural flora, quite apart from pollination of garden and farm produce.

7. Dust treatment is said to work with good effect against varroa mites. Why has this safe and cheap method of control not been included in the new DARG booklet on varroa?

We did not include dust in our new DARG varroa booklet because we were unable to trace any effective use of it in any country. Even anecdotal evidence proved to be second hand when investigated.

8. The new Ministry Leaflet still describes the tobacco smoke method for diagnosing varroa infestations. Is this method really safe in the light of reports of severe bee losses when using tobacco?

As shown in 5 (b) we have found tobacco smoke relatively ineffective. In my lectures and demonstrations in and out of Devon I have felt obliged to include it, but heard many reports of distressed bees. I have always stressed the uncapping of drone brood in early April as being a superior technique. In my experience most beekeepers on the continent abandoned tobacco smoke at least two years ago.

VARROA:
VARROA MITE KNOCK-DOWN RATES

BKQ 32
Winter 1992/3

The raw figures of Brian Cant's hives are quoted (Hive F). It may interest readers to apply parallel reasoning to arrive at the initial infestation here.

Experimental work by Brian Cant on single brood chamber National hives in South Devon, with daily count of mite knock-down on paper inserts.

Assumptions, deductions and calculations by Ron Brown, at whom all criticisms should be aimed!

Deductions from Data (Hive G)

1. The first plateau equated to steady knock-down of mites

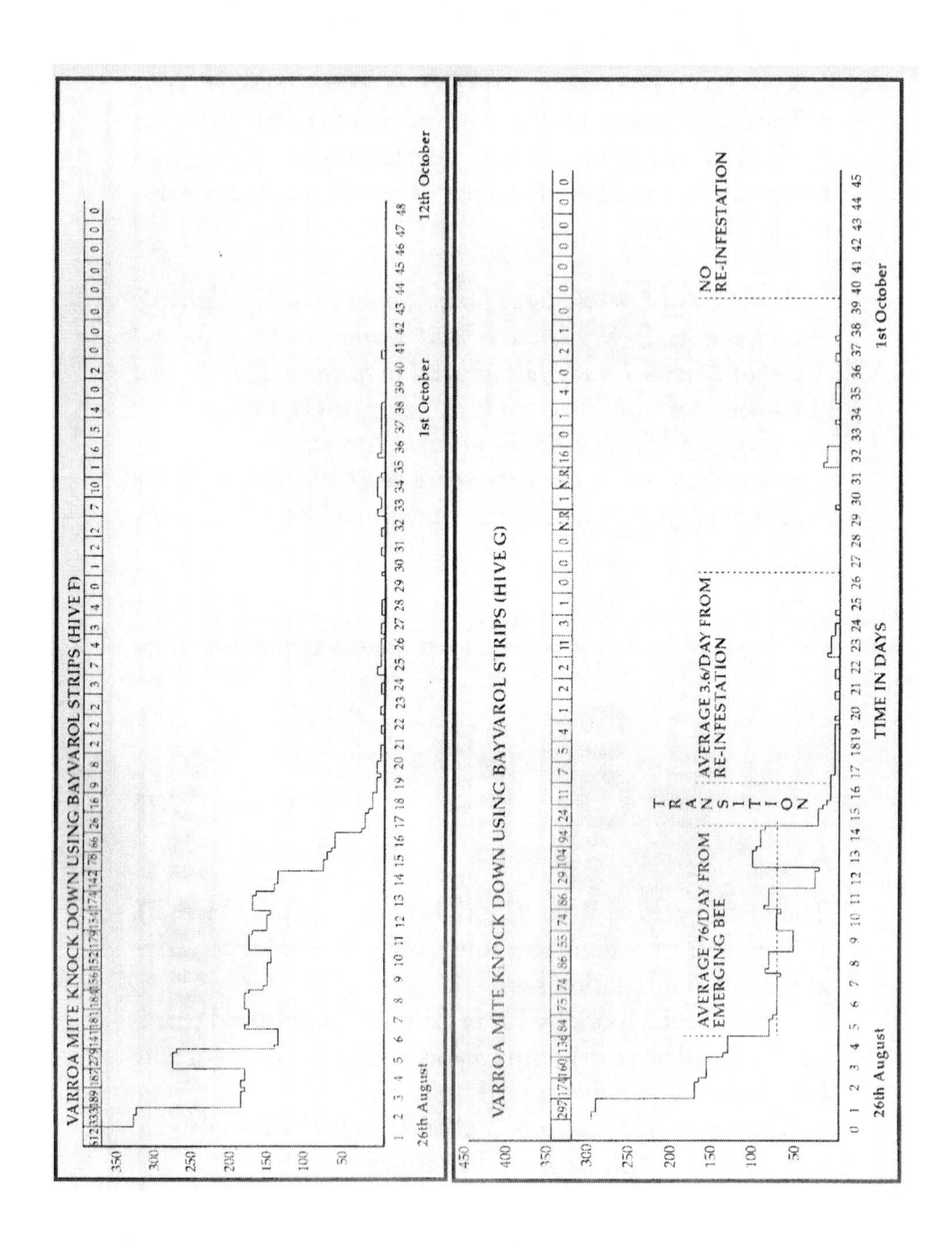

VARROA MITE KNOCK DOWN USING BAYVAROL STRIPS (HIVE F)

VARROA MITE KNOCK DOWN USING BAYVAROL STRIPS (HIVE G)

off newly emerging bees - about 76 per day.
2. The second plateau - steady knock-down by re-infestation.
3. Final plateau suggests absence of re-infestation (rain, bad flying weather).
4. Total infestation 25th August - 823 mites.
5. Initial 'circulation' period for Bayvarol distribution by bees - 4 or 5 days.

N.B. Natural mortality before treatment with Bayvarol was 4 mites per day.

Assumptions (Hive G)

1. That re-infestation was constant at 3.6 mites per day over the first 16 days, as in next 10 days.
2. That C. 50% mother mites emerge from brood cells plus 1.5 mated daughters.
3. That from 26th August on, only worker brood involved.

Calculations (Hive G)

Ignoring immature mites in cells initial infestation was:
(460-14) + 1/2 (760 + 30 - 36) 446 + 377 823 (Day 0) (on mature bees) (in cells) i.e c. x200 of natural mortality (4/day).

VARROA:
DEVELOPMENT OF VARROA MITE RESISTANCE TO SYNTHETIC PYRETHROIDS

BKQ 49
Spring 1997

As so often happens, rumours spread faster than facts, and many beekeepers in Britain fear that the resistance of varroa mites to synthetic pyrethroids is much more widespread than it actually is. Our Devon group which visits the continent each summer was pleasantly surprised to find last year that knowledgeable beekeepers in Alsace (E. France) as well as in S. Germany, Freiburg area, were still using apistan with complete confidence. As recent investigations by the French authorities have clearly shown, resistant mites are still found only in the south, except for two incidents clearly diagnosed as having been spread by beekeepers. The geographical distribution seems to confirm that the strain originated in Sicily, spread to Italy and thence into SE France. The pattern of varroa resistance suggests that it arises from a specific strain of mite rather than a series of spontaneous outbreaks ever a wide area. The isolated case in Maineet-Loire (NW area) was proved to have arisen from regular importation of queens plus workers from Italy, and cases east of Lyon correspond to transfers of hives from the SE.

Apart from these, it seems that there were two main avenues along which the resistant strain moved, (a) along the valley of the Gavonne, (b) going up the valley of the Saone, The Massif Central as well as the north, the east and the west appear to remain free of the resistant strain. The actual progress closely resembles that of the original varroa infestation in the first place, though in different areas; maybe even slower as beekeepers are now more aware of the problem.

Following upon clear evidence of mites found to be resistant to fluvalinate, first in Sicily and later in other parts of Italy, French authorities found no evidence of any such resistance in France until late in 1995, when five apiaries close to the Italian border in the SW were detected, after thorough checks on the efficiency of apistan treatment.

As far as I know, no examples of mites resistant to Bayvarol/flumethrin have yet been found, but as the chemicals in both Apistan and Bayvarol have certain similarities (e.g. a fluorine atom, a cyano group), it seems likely that resistance to either could rapidly be followed by resistance to both. Probably this is why the French have licensed the use of Apivar/amitraz as the official alternative, also as slow release strips used in an identical manner. On the assumption that the resistant strain is spreading slowly up from the south, it may be three or four years before it reaches the Channel coast, and a year or two after that before it reaches Britain, like the mite did a few years ago. This is subject to two points: that we do not freely import bees from the continent, and that new, local resistant strains do not develop locally because of our carelessness in treatment.

VARROA: THYMOL V VARROA

BKQ 60
Spring 2000

Many beekeepers are using thymol as a 'soft' option for controlling varroa mites in their apiaries. Ron Brown has used thymol and carried out many interesting experiments with the substance and in the notes below we have a record of his findings and experiences over the last few years as well as a wealth of miscellaneous information on this useful 'apiculural' substance. Editor

1. Thymol

- is supplied in the form of small white crystals which are very soluble in alcohol, but almost insoluble in water (only 1 part in 1600). It stays in solution when relatively small quantities of that solution (in alcohol) is added to bulk sugar syrup.

It is harmless to *Apis mellifera*, presumably because our bees have been foraging on thyme plants in Europe and the Mediterranean Basin for millions of years and they have probably developed a complete immunity to any original toxic effect. It is toxic to varroa mites as they originated in the Far East and possibly because there are no significant amounts of thyme found there.

2. What is Thymol?

Chemically thymol is 3-Hydroxy-p-cymene, originally made by steam distillation of thyme plants to obtain oil of thymol, but now manufactured in chemical factories by the oxidation of piperitone, which is contained in eucalyptus oil. The crystals have a melting point of 51.5 degrees C, a boiling point of 233.5 C, and a molecular weight of I5O. Empirical formulae is C1 OH14 0 and a structural form as shown:

3. Composition of Italian Sachets of Apilife Var:

* 74% thymol
* 17% oil of eucalyptus
* 4% camphor
* 3% menthol
  2% vermiculite

4. Thymol Dispersion

At normal temperatures and pressure thymol crystals are well within the solid phase; only above 50 degrees C would the liquid phase be involved. At 0 degrees C vaporisation would be very slow and very difficult to measure. At 30 to 35 degrees C (broodnest temperatures) thymol sublimes from solid to gaseous phase, the actual speed of sublimation (vaporisation) increasing rapidly as Melting Point is approached. For practical reasons the thymol crystals must be confined within a container having very fine apertures, like fine denier artificial silk, for example, acting as a semi-permeable membrane analogous to organic membranes, but at a different order of magnitude. As seen below, at brood nest temperatures, 4 gm of thymol crystals would be almost completely vaporised within about 80 days.

So far as thymol residues in honey are concerned, this is unlikely to be harmful as thymol is commonly used in cough and throat medicines. The practical Swiss use the term 'level of customer perception' which they have found to be well above any residues.

5. Vaporisation of Thymol Crystals

Several years ago I set up a wooden box with a small electrical heater (controlled) and adjusted conditions in a sealed room until the temperature in the centre of the box was 32 C / 90 F +/- 5 F (brood nest temperature). A watch glass of white thymol was weighed every 24 hours and the observations in Table 1 were made.

6. Use in Winter Feed

Thymol was used for many years by ROB Manley, ostensibly to preserve sugar syrup; coincidentally, despite keeping hundreds of stocks for many years, he wrote in his Beekeeping in Britain, 1948, that he had no experience of nosema (pp 269 & 354)). On a much smaller scale, I have kept 20 or more stocks for over 30 years with thymol in winter syrup (and when necessary in spring), with no trace of nosema. I use slightly larger amounts than Manley did, dissolving 40g thymol crystals in 200ml of surgical spirit to make up a

Table 1		
original weight of thymol	4.374g	
After 1 day	4.305g	loss in weight 69mg
2 days	4.225g	50mg
3 days	4.204g	51mg

ie within expected experimental error a reasonably concordant result of 57mg per day, c. 1.3% of median weight per day.

stock solution of 200g per litre. Of this 4ml (a teaspoonful) is added to every 5 litres (or gallon) of syrup containing 4 kg (9 lb) of sugar when given to bees as their winter food in mid-September.

The cost: Thymol (in 5 kg lots or more) @ £19 per kg from Treat & Co, Northern Way, Bury St Edmunds, Suffolk (tel. 01284 702500). Surgical spirit @ 95p for 200 ml - from any chemist. Total cost per colony per year - roughly, 30p.

7. Use in Summer

One full teaspoon (5 g) in a recycled tea bag, sealed with tape. Two such sachets are placed on top of brood frames in April. They do not need to be removed as the bees either propolise them down or throw out the empty sachets when all the thymol has vaporised - usually by the end of July. Cost per colony - about 20p per year.

8. Some Practical Details

(i) Accidental overdose

Four years ago, in May, at an outstanding apiary up in the Teign Valley:

At this time I was still using bags made up from old ladies' tights (used tights, not taken from 'old' ladies). I carelessly snagged one over the top of a crowded brood box and 8g of thymol crystals fell out onto top bars and seams of bees. I closed up the hive and went home a very worried man. Two days later I returned expecting to find evidence of a disaster but found not a single dead bee and no trace of crystals on any top bar; the bees were flying busily and happily. Later on, I artificially swarmed that hive, made two four-frame nucs from the upper old brood box and finally harvested three heavy supers of honey in the second half of August.

(ii) Eucalyptus Oil (Apilife Var, 17% eucalyptus)

Three years ago in July I had to lecture and demonstrate anti-varroa techniques at an apiary on a farm in Berkshire. I lectured in the barn during the morning and after lunch, on the hottest day of the year (the ambient temperature at 3.00 pm being over 30C, nearly 90F), I walked towards the apiary with my Berkshire friends. We could smell eucalyptus from 25 yards away and in the beeyard all was chaotic. There were bees hanging in skeins from hive roofs and floors, knots of bees scattered in the grass (some even with queens). I took off hive roofs and shook bees back in, plus scraping some up from the adjacent grass, and I left the hives with roofs diagonally on top to air, having removed packs of Apilife Var. Three hives had not been given this treatment and were flying normally. These were given a hand spray of dilute lactic acid, avoiding frames with unsealed brood. The packs did carry a warning about use in very hot weather, but this had been thought of in relation to Italy, not Berkshire.

(iii) Visit to Liebefeld Research Labs - 1995

Quite by accident I was chatting to a Swiss lady worker in her corner of the main lab and asked what she was doing. "Working on LD50 values for various substances used in hives against varroa." "That one smells like eucalyptus!", I exclaimed. "Yes. and it is very bad for bees. The LD 50 value is far too close to the amount used in some treatments." I have never seen any official report of her work and apologise for what may be a breach of confidentiality, but after several years I may be forgiven. I most certainly, though, will not ever contemplate having eucalyptus in any of my hives.

9. My Current Practice

Appreciating that varroa mites increase most rapidly on drone brood during summer months, I use thymol sachets in summer, and thymol in feeding syrup in autumn and winter. So, my hives have thymol vapour during summer, which probably masks the pheromone given off by the larve about to be capped in their cells and reduces the number of mother mites finding their way in. It is also directly lethal to the mites exposed to it. In autumn and winter bees have some thymol in their stomachs which may reduce the effect of nosema organisms, as well as being hostile to varroa mites, if only to a limited extent. I also put in Bayvarol or Apistan strips in early September.

SECTION 6.
BROTHER ADAM, AN APPRECIATION

BKQ 47
Autumn 1996

By the death of Brother Adam the world has lost a great, perhaps its greatest beekeeper, but we in Devon have lost a great friend, and our Life President. German born but having lived as our close neighbour at Buckfast since 1910, Adam has spent more years in Devon than most of us. At our very first residential beekeepers' weekend course at Seale Hayne in 1929 he was there, lecturing alongside William Herrod-Hempsall. Sixty years later he was still attending, but needing some help down those steep steps to the dining hall. For year after year he attended our County AGM in Exeter and honoured successful candidates by presenting their BBKA Exam. certificates. It was sometimes my privilege to collect and take him, and remember his hatred of seat belts; typically he would never wear one until it was a legal requirement. More than once we met by chance near Grimspound on Dartmoor, where my heather site was about a mile from his. He was greatly amused by my wheelbarrow on the roof rack of an ancient Maxi containing four National hives.

About 16 years ago Owen Meyer (then BBKA General Sec) and I interviewed Adam and the Abbot to express concern about the long-term future of this unique bee enterprise. Our suggestion that a young beekeeping Brother might be transferred to Buckfast from a continental monastery was laughed at, as apparently such a procedure was unheard of.

At an interview on Adam's 90th birthday Peter Rosenfeld and I renewed the suggestion, but clearly Adam felt that all the long-term work had been done and he said that Peter Donovan was well able to continue on the lines established. Probably any attempt to replace a Da Vinci, a Newton or an Einstein would also have been doomed to failure anyway.

Eighty years ago Adam was considered not strong enough to continue as a stone mason helping to build the Abbey, and was transferred to help Brother Columban with 40 hives of bees, and in kitchen work. Fifty years ago gross overwork caused a break-down in his health for several months, also a heart condition which doctors said would prevent any further work. Somewhat similar advice was also given twenty years ago, yet in his nineties Adam was still working in his out-apiaries, and being seen by millions on television. What an example of will power triumphing over every obstacle. His last years were spent in a comfortable retirement home about a mile and a half from the Abbey. Some of us visited him occasionally and usually took a bottle of German wine. I did not dare to take a bottle of my mead! Memories of enjoying his vintage heather mead, a glorious golden mahogany colour, in his bee sanctum were still too vivid, and far beyond my amateur efforts. For a year or two we managed to lift him round to a local hotel for our annual DARG lunch and General Meeting, but last year even this was not possible. As his Abbot said "May he rest in peace."

Brother Adam

SECTION 7.
THE KILLER BEES OF BRAZIL

The Beekeepers Annual
1983

Background

From time to time we read horrifying accounts and see frightening films of bees alleged to be so dangerous that they might almost threaten mankind at large. Where does the truth lie? As one might expect, it is not nearly so bad as reports in the media suggest, but it is serious, at least to beekeepers. Here are the facts: honey bees were introduced into Brazil by the Portuguese in the 18th century, but were never very successful; despite a favourable environment and the vast areas available, total annual honey production was still under 5,000 tons in 1950. With the intention of breeding better bees, 28 queens were imported from Taborah (Tanzania) and Pretoria in 1957, the idea being that the climatic conditions were roughly equivalent to those in Brazil.

Accidental Queen Release

In order to prevent the escape of swarms, the hives with African queens (at Rio Claro research station in southern Brazil) had Queen Excluder strips across the entrances. However, a well-intentioned beekeeper, unaware of the experiment in progress, noticed pollen loads being lost at the congested entrance and pulled off the Queen Excluder strips. Before anyone realised what had been done, 26 swarms had escaped in two or three days, headed by African queens. For a year or two nothing untoward occurred, and then reports of very aggressive bees started to come in over an area of 50 or 60 square miles. By 1963 Adansonii bees were reported in an area 500 miles across, and since then have been advancing by 150 to 250 miles a year, reaching Argentina and Guyana by 1975, Venezuela in 1976 and Panama in April 1982. At this rate they will be in Mexico before 1986 and a real threat to the U.S.A. by 1990. At first many beekeepers in Brazil gave up and stocks were burned; local authorities banned beekeeping in or near urban areas.

The Beekeepers Annual 1983

The Florianopolis Conference

In 1978 there was a conference on the problem in Florianopolis, Santa Catarina Province, Southern Brazil. This was attended by hundreds of beekeepers from all the S. American states, plus about 20 from the U.S.A. and another dozen or so from tropical areas in other parts of the world. I was invited, and apart from a freelance journalist, was the only one from the British Isles.

No-one attended from France or Belgium, one from Portugal and Dr Koniger from Germany; the presence of a contingent of Venezuelan fireman (in full uniform), underlined the seriousness of the situation!

For 3 or 4 days papers were read by well-known beekeeping

scientists and by civil servants in the Agriculture Departments of various S. American countries. My humble contribution (later published by Apimondia) was a paper on "Adansonii Management" based on 12 years beekeeping experience in Central Africa, most of the time with six to ten hives in my garden in Lusaka and Ndola. getting over 100 lb a year of honey per hive regularly without worrying my neighbours on either side, using techniques developed over several years. In these first few years. on various occasions our chickens were all killed, we were besieged by angry bees for hours at a time and so on, before I learned by trial and error how to operate.

Alas, the Brazilians were more impressed by the American approach, dealing with very large scale operations. My plea that 'small is beautiful and 'a cash crop for 100,000 peasants is better than 100 huge honey farms' was not taken seriously. All the same, when I spoke I noticed my audience increasing rapidly after the first minute or two until there was standing room only (loudspeakers outside sent the proceedings into every room and lobby - we were using the Provincial Parliament buildings). Was this to hear the great message on how to cope with African bees? No, although I was indeed flattered to be told why they all came in. It was, they told me later, to see who was speaking such fluent English in such a clear voice! Gratifying, in this case, to have been the only Englishman present!

The Tour

After these sessions, we visiting beekeepers were taken on a comprehensive tour of apiaries in Santa Catarina and Parana provinces, over many hundreds of miles, staying in different towns, sometimes flying, sometimes by coach. The hospitality was out of this world and the tour quite unforgettable. In general terms their approach to the problem was to site large apiaries in forest areas remote from towns, with hives on single stands (to prevent vibration exciting other stocks), using smoke on a huge scale. Their smokers were the size of

Full protective clothing and large amounts of smoke are needed when handling Adansonii bees in the forest apiaries of Brazil

small dustbins, and the volume of smoke comparable to that used to cover infantry attack in war.

I next spoke to Helmut Wiese (President of the Confederation of Brazilian Beekeepers) in Mexico last October (1981) and he told me that honey yields had greatly increased and were now up to 15,000 tons a year (Mexico and China at about 50,000 tons each are at the top of the league table). So the original purpose has been achieved, at least, but the threat to urban areas by wild colonies persists, and it is sad to realise that honeybees are pests to be destroyed when found in wild colonies in or near towns.

Why so Aggressive?

Why is the Adansonii bee so aggressive and so prone to swarm frequently and over such large distances? I think these characteristics have been programmed into them in the following manner. Africans were (and largely still are) bee hunters rather than beekeepers, and if over tens of thousands of years one kills and robs out the 'soft targets' and leaves alone the more vicious ones, then gentle bees are eliminated and those that remain are genetically 'selected' for aggressiveness. The unusual swarming propensity can be accounted for by the shortage of reasonably sized cavities. The trees are smaller because of frequent bush fires over most of the area in the long dry season, and I frequently found colonies of bees occupying cavities that no European bees would have been interested in, even in street lamp posts. The need for swarms to travel long distances also arises from frequent fire danger - only colonies flying 20 to 30 miles could get out of the burning area. About the distances travelled there is certainly no doubt; many times in September and October I have heard the noise of a swarm coming from a distance, like a small helicopter, flying overhead in a whirling cloud of bees at a speed of 10 to 15 m.p.h. and continuing out of sight, although on occasions followed 5 to 10 miles before losing contact. So all these factors:- frequent swarming, swarms travelling long distances, an aggressive nature, have become well established genetically. The hostile environment (including hostile humans) has permitted only bees with these characteristics to survive.

What of the Future?

The hope that by hybridising with local bees the aggressive strain would be diminished has not so far been realised. In 1978 the bees I saw and photographed in Brazil were just like my bees in Lusaka and Ndola; slightly smaller than those in Europe, very aggressive and good honey-getters. In my opinion they present no threat to temperate climates. Nearly 30 years ago, working in Lusaka, I established their threshold working temperature at 57°F. Over many days in our Central African 'winter' I correlated a count of bee traffic at hive entrances with ambient temperature and below 56°F found virtually no activity. On a cloudy day in July (at 4,000ft a.s.l.), there would sometimes be a maximum temperature of less than 57°F and then scarcely a bee flew. On most days the temperature reached 57°F before midday (often much more) and the change in flight activity between 56°F and 58°F was dramatic, both in commencing and in terminating as the temperature fluctuated. At home in Torquay my bees in March behave similarly, but over a temperature range about 10° cooler (except for water-carrying, which goes on at temperatures down to 41°F or 42°F).

I have no intention of introducing Adansonii to Torquay, but if I did, their spring build-up would be so slow, that they would be many weeks behind our bees and unable to compete. A threat to Mexico and the Southern States of the U.S.A., yes. To most of North America - No!

SECTION 8.
RON BROWN OBE BSC

BKQ 104
Summer 2011

We were very saddened to hear of the passing of Ron Brown, someone who had been mentioned often bee-orientated household, although his wife, Rosemary, had kept in touch with us and we had known of his increasing frailty. The following paragraphs were first published in BKQ 86, in the Bookshelf Special for that issue, and we make no apology for including them here as a reminder of his exciting, varied and eminently useful life which culminated in the new edition of his "Beekeeping - A Seasonal Guide", which fortunately, Ron was able to see and approve of just before he died. John was involved in the editing and bringing up to date of the text.

THE MAN AND HIS BOOKS AN APPRECIATION
by Val Phipps

BKQ 86
November 2006

Though the name of Ron Brown has been mentioned often in our bee-orientated household, my interest in this many-faceted, idiosyncratic and humorous man did not develop until I picked up his two autobiographical works, attracted intiially by their unusual jacket designs.

The "Prof" as he was then known, was, before the war, station Education Officer at RAF Bircham Newton, in west Norfolk. He became station Intelligence Officer at the start of the war and an expert in communications of all kinds plus radar, the allies secret weapon. Influenced by stories I had heard during my teacher-in-training years at Bletchley Park College, I imagined that the "Prof" would have been chained to a desk amid conditions of complete secrecy - far from the truth. His war was an extremely active one and his accounts of his adventures make riveting reading.

The "Prof's" memory of his pre-war years in RAF Coastal Command remains very detailed; his life was hectic and interesting, with a very good social life. It was also very adventurous - his love of adventure followed him throughout the war and into his subsequent career in Africa.

After varied duties in Coastal Command, including the taking of aerial photographs, he was transferred to Iceland, operating ice reconnaisance patrols, flying as observer and photographer far north of Iceland, and visiting Greenland, before being stationed at Gibraltar between 1943 and 1945. During the war he was four times mentioned in dispatches and awarded the OBE.

When the war ended, Ron Brown went, with his first wife, Carol, to Africa where he taught Maths at Durban High, then Science in Broken Hill and Lusaka, eventually becoming Education Director in Northern Rhodesia (now Zambia) from 1959 to 1964. It was whilst living in Lusaka that a chance swarm, deciding to take up residence in his garage, initiated his lifelong fascination for bees. Watching their behaviour and questioning aspects of it rapidly became almost an obsession, which was fuelled by a lack of satisfactory answers from books or colleagues and which developed into a philosophy of never believing anything until he had tried it himself.

Life in Africa was endlessly fascinating - coping with insect life and snakes which liked to wrap themselves round the front axle of a slow moving car; stopping his children from stroking the crocodiles whilst on a lake in a dug-out canoe - balanced by the demands of a working life in a country where as a Science teacher he had to pretend to be a white witch to impress his pupils! His skills were further developed when, having persuaded officials to build a swimming pool to avoid the ever-present problems of Biharzia in local streams and rivers, he had to become chief fund-raiser, designer and clerk of works!

Ron Brown's enthusiasm for bees developed further when he retired to Devon in England and became editor of his county's monthly "Beekeeping" magazine. As his experience and knowledge grew over the years, he became well-known internationally as an excellent lecturer. He has been on radio and television, informing the general public more recently of the problems caused by Varroa destructor. He is a former President of Devon Beekeepers' Association, has been chief examiner for the University of Cambridge Examinations Syndicate and, besides his wonderful autographical books, "All around the compass" and "Ex Africa", he has written several important books on a variety of beekeeping topics, all of which are featured here.

KEY DATES

April 1953

Lusaka, Central Africa, took his first swarm of bees, continued to keep bees in Africa, single-handed until 1963.

May 1965

Secretary of Torbay Beekeepers Association, Devon UK.

1975 -1982

Editor of "Beekeeping".

1981

"Beeswax" won the Silver Medal at the XXVIII International Apimondia Congress, Acapulco, Mexico.

Also: Founder member of the Devon Apicultural Research Group (DARG), and President since the death of Brother Adam. Chairman of Torbay BKA.

President of Torbay BKA - since the death of Brother Adam.

THE BOOKS
AUTOBIOGRAPHICAL BOOKS:

"All Round the Compass with RAF Coastal Command"

Janus Publishing Company, 1993. Hardback, 136 pages, illustrated with B&W Plates. ISBN 1-85756-081-7. Excellent pictures of the "Queen Mary", "Swordfish", "Condor FW200" "Beaufighter" and "'Hoopla Wellington" airplanes, pack-ice, de-icing the "HMS Cumberland" during an Arctic Convoy, US Navy airships, and the author's draw-ings of maps. A fascinating study of one man's war years.

"Ex Africa"

Published 1996, Venture Press, Torquay, Paperback, 112 pages, illustrated with colour and B&W plates, and diagrams. ISB~ 0-9528936-0-6.

After leaving his own country for a career overseas. Ron's stories about these years show how he and his family became fully integrated in their new surroundings and took advantage of all the opportunities which came their way, thus enriching their own lives and those around them.

BEEKEEPING BOOKS

Ron Brown's books on beekeeping can be divided into two distinct groups - those concerned with practical beekeeping and those which deal with specific aspects of beekeeping history.

However, probably his most successful book, his award-winning "Beeswax" (Published 1981. Paperback 87 pages. Illustrated with B&W plates and line drawings. Third edition, 1995 Bee Books New & Old. ISBN 0-905652-36-3) is a hybrid of the two. The first chapter gives a survey of the uses of wax from classical times in Rome up to the rebuilding of parts of St Paul's Cathedral and the Wax Chandler's Hall in London after the 1940 air raids. Interestingly, both of these buildings had been previously destroyed some three centuries earlier during the Great Fire of London. After a detailed section on the origins of beeswax involving a look at bee anatomy and physiology, Ron turns to the more practical side of things: how the beekeeper can render and use the beeswax in the home as an essential ingredient of numerous products, complete with recipes and technical information. The process of making wax foundation for one's bees is also described and as well as producing excellent quality wax for showing. That the book has run to three editions and won a prestigious Apimondia Silver Medal shows how well-respected and popular this small but important volume has become.

Ron's beekeeping guides "Beekeeping - A seasonal guide" (Batsford 1985. Hardback, 195 pages. Illustrated with B&W plates and line drawings. ISBNO-7134- 4489-4) and "Honey Bees - A guide to their management" (Crowood Press, 1988. Hardback 128 pages. Illustrated with B&W plates and line drawings. ISBNl-85223- 051-7) are both obviously the work of an experienced teacher who can clearly describe, step-by-step, the important management techniques needed to keep bees. The first book takes the reader season by season through the beekeeping year explaining exactly what needs to be done and when, and gives advice and instructions for beekeepers of different levels of experience. The second book is a much more straightforward account of beekeeping practice, again mainly season by season, though not on the same scale as the former work. Both books are filled with many of Ron's delightful sketches which help to clarify points raised in the text. Ron writes from many years of experience and gives his readers helpful beekeeping tips, not found in any other books, which he himself has devised for the successful management of his own bees.

Primarily a book written for Devon beekeepers "One Thousand Years of Devon Beekeeping" (Published in March 1975 by the Devon Beekeepers Association. Paperback, 66 pages) is, in essence, a history of beekeeping in England and thus it has more than local appeal. The reason for this is that Devon was the stamping ground of many worthy beekeepers who helped to shape the craft over the centuries including the Rev Jacob Isaac (the 'moving spirit' behind the Western Apiary Society) and Thomas Woodbury (inventor of the Woodbury hive and importer of Italian queens). The second part of the book concerns the development of the DBKA and mentions the parts played by most of the key players during the years 1875 - 1975 as well as notes from the minutes and items on visiting lecturers (eg W Herrod-Hempsall). Part of the book concentrates on the association's journal "Beekeeping" which contains a wealth of interesting anecdotes and makes me, for one, long to read past pieces of Rev Hustwayte's 'Diary of Samuel Pepys, Beekeeper' or experience the dry humour of Captain Turner's "old Drone". Interesting too, is the suggestion put in 1947 to the committee that DBKA should discontinue their own journal since Bee Craft had then been adopted by the BBKA as its official journal. However, 'by a very large majority' those present voted to continue with their own journal. A special small chapter is devoted to the apiary at Buckfast Abbey and the forward to the book which marked the centenary of the DBKA is by Brother Adam himself.

"Great Masters of Beekeeping" (Published by Bee Books New & Old, 1995. Hardback 110 pages. Illustrated with B&W plates and drawings from early books. ISBN 0-905-652-31-2) could almost be described as a sequel to '1000 years..' but on a larger and grander scale. Here Ron Brown explores the lives and describes the contributions of many beekeepers, both national and international, who have helped shape beekeeping as we know it today. These include Charles Butler, Rev John Wilkins, Rev John Thorley, Anton Jaska, Thomas Wildman, Rev J Isaac, Francis Huber, Dr Edward Bevan, Rev L Langstroth, Rev J Dzierzon, Thomas Woodbury, Charles Nash Abbott, Johannes Mehring, Franz von Hruschka, Abbe Collin, T F Bingham, E C Porter, W B Carr, Thomas Cowan, Rev J G Digges, J and W Herrod-Hempsall, L Snelgrove, Annie Betts, Karl von Frisch, ROB Manley, E and A Alphandery, Dorothy Hodges, Colin Butler, Norman Rice, and, of course, Rev Brother Adam. Indeed, a whole galaxy of beekeeping stars! Excellently-researched, a fascinating read and well worth getting hold of a copy.

Lightning Source UK Ltd.
Milton Keynes UK
UKOW03f0911060215

245751UK00002B/12/P